MITIGATION OF LANDFILL GAS EMISSIONS

Mitigation of Landfill Gas Emissions

Małgorzata Pawłowska

Faculty of Environmental Engineering, Lublin University of Technology, Lublin, Poland

CRC Press
Taylor & Francis Group
Boca Raton   London   New York   Leiden

CRC Press is an imprint of the
Taylor & Francis Group, an **informa** business

A BALKEMA BOOK

CRC Press/Balkema is an imprint of the Taylor & Francis Group, an informa business

© 2014 Taylor & Francis Group, London, UK

Typeset by V Publishing Solutions Pvt Ltd., Chennai, India
Printed and bound in Great Britain by CPI Group (UK) Ltd, Croydon, CR0 4YY

Published by: CRC Press/Balkema
   P.O. Box 11320, 2301 EH Leiden, The Netherlands
   e-mail: Pub.NL@taylorandfrancis.com
   www.crcpress.com – www.taylorandfrancis.com

ISBN: 978-0-415-63077-1 (Hbk)
ISBN: 978-1-315-75533-5 (eBook PDF)

Table of contents

Preface

Landfilling was and still remains an important component of municipal solid waste management. Despite the enormous progress that has already taken place in terms of reducing its role in the waste management, landfills are still needed in the modern world. Landfilling is the most cost efficient method of waste disposal, which explains its prevalence. Unfortunately, its low cost is not connected with a low environmental risk. Deposition of waste in the landfill poses a threat to the purity of water, soil, and especially air. What is more, due to the slow rate of decomposition of organic matter under anaerobic conditions negative effects of the landfill are becoming problematic for future generations.

Globally, the use of other methods of municipal waste disposal, such as thermal treatment, is not popular. The high cost associated with the purchase of technology, construction of the plant, professional service and control, as well as the transport of waste from large distances constitute the main barriers. Many countries will not be able to afford to invest in new expensive technologies for waste treatment for a long time. The choice of methods of waste disposal often also depends on social issues. The presence of a waste incineration plant in a neighborhood usually raises local residents' objections.

In the coming years, a radical decline in the share of landfilling in the global waste management should not be expected. However, it does not mean that people are powerless in the face of the emission of harmful gases into the atmosphere, the spread of bioaerosols and odors. There are many ways of preventing or protecting the environment against the negative impact of landfills. Many of them requires the involvement of substantial financial resources, therefore, their use in practice is limited. However, there are also other solutions that can contribute to a significant reduction in emissions from landfills at low cost.

In this book, several ways to reduce the negative impact of landfill on the environment are presented. Consecutive chapters presenting the concepts of anaerobic bioreactor, aerobic and semi-aerobic bioreactors, deal with possible influences on gas emissions by the modification of the conditions prevailing in the waste deposit. The last chapter raises the issue of efficiency of microbiological oxidation processes, which remove methane and trace gases from landfill gas in different types of biotic systems, such as biocovers, biofilters, or biowindows.

Firstly, the problem of waste disposal was highlighted from the global perspective, paying attention to the role that it plays in the waste management in different parts of the world, and the effects of its deposition; with particular emphasis on greenhouse gas emissions into the atmosphere. Furthermore, a separate chapter is devoted to the characteristics of the landfill gas. Its composition and properties, which determine the nature of the threats to the environment and potential ways to reduce its emissions, are described.

It should be borne in mind that the present work does not exhaust the subjects and further examination is required. The author's aim was to show the main directions of the development of methods to reduce the impact of landfills on the environments, which appear in different parts of the world. The study focuses on the use of biological processes, whereas issues related to technical protection landfills, such as sealing the basin, more efficient systems for landfill gas extraction and method of its disposal were omitted.

Finally, I would like to express my appreciation to Ms. Katarzyna Wszoła for her invaluable help in preparation of this book.

About the author

 Małgorzata Pawłowska, Ph.D., Sc.D. is a researcher and lecturer working at the Faculty of Environmental Engineering of Lublin University of Technology. From 2012 she is a head of the Department of Alternative Fuels Engineering.

She received her M.Sc. of the philosophy of nature and the protection of the environment at the Catholic University of Lublin in 1993. In 1999 she received her Ph.D. in agrophysics after defence the thesis entitled: "A possibility of the reduction of methane emission form landfill by its biochemical oxidation in landfill cover—model study", at the Institute of Agrophysics of the Polish Academy of Science.

Mrs. Pawłowska is an author and co-author of fifty five papers and book chapters, two monographs, one handbook on biofuels, five patents and several patent applications. She is also the co-editor of two monographs.

The major areas of her scientific interest is control of gas emission from landfills, with special attention to methane and BTEX biooxidation in landfill covers and biofilters. She is also working on biogas production from waste.

Abbreviations and symbols

ABL	- anaerobic bioreactor landfill
BOD	- biochemical oxygen demand
BMP	- biochemical methane potential
CFU	- colony forming unit
COD	- chemical oxygen demand
COS	- carbonyl sulphide
DCM	- dichloromethane
DMS	- dimethyl sulfide
DMDS	- dimethyl disulfide
d.w.	- dry weight
EBRT	- empty bed residence time
EC	- elimination capacity
EPS	- extracellular polymeric substances
GHG	- greenhouse gases
GDP	- gross domestic product
GWP	- global warming potential
HRT	- hydraulic retention time
K_M	- Michaelis constant
K_S	- saturated hydraulic conductivity
MBT	- mechanical-biological treatment
MMO	- methane monooxygenase
MN	- methane number
MSOR	- mechanically sorted organic residues
MSW	- municipal solids waste
NMVOCs	- non-methane volatile organic compounds
LHV	- lower heating value
LFG	- landfill gas
OLR	- organic loading rate
OFMSW	- organic fraction of municipal solids waste
PCE	- tetrachloroethylene
pMMO	- particulate methane monooxygenase
PTAFBL	- pump and treat aerobic flushing bioreactor landfill
Ptotal	- total phosphorus
RE	- removal efficiency
RMSW	- residual municipal solid waste
RuMP	- ribulose monophosphate cycle
sMMO	- soluble methane monooxygenase
TBE	- trichloroethylene
TeCA	- tetrachloroethane
TOC	- total organic carbon

TN   - total nitrogen
TKN   - total Kjeldahl nitrogen
TS   - total solids
V_{max} - maximum activity
VOCs   - volatile organic compounds
VS   - volatile solids
VS_{add} - volatile solid added

CHAPTER 1

Landfilling of municipal solid waste in global perspective

1.1 INTRODUCTION

Landfilling is the oldest and the simplest form of waste disposal. Over the centuries the process has gone through significant changes, from the form posing a serious threat to the environment—such as filling naturally occurring repositories or anthropogenic excavations in the ground (open dumps)—to ending as engineering projects, properly designed with a view to reduce the impact on the environment (sanitary landfill). However, it still has a negative impact on water, air and soil, creating hazards on both a local and global level. These sites occupy large areas of land, which are excluded from use for other purposes for a long time. Moreover they pose a threat to the health and lives of people, and their impact is long-term.

Currently, apart from the primitive open dumps, landfills are classified into three groups: 1) semi-controlled and controlled dumps without liners and systems of landfill gas and leachate management, 2) engineered/controlled landfills with special technical and operational solutions minimizing the landfill impacts on the environment, such as liners, leachate collection and treatment, passive or active degassing system, waste compaction, daily cover, water monitoring, etc., and 3) sanitary landfills with technical and operational conditions, as in the case of controlled landfill, but having an additional post-closure plan, more advanced leachate treatment and LFG burning with and without energy recovery (Hoornweg & Bhada-Tata, 2012). Despite the undeniable progress of civilization, it is possible to find all the aforementioned forms of waste storage, including open dumps. What form dominates depends primarily on the level of economic development of the country. In developed countries, equipped with not only the right "know-how", but also adequate financial means special attention is paid to the conditions of waste deposition, providing proper isolation of the waste mass and control throughout each phase of the landfill biochemical evolution. In less developed countries, environmental problems are pushed into the background, to give priority to the basic needs of the population, such as creating decent living conditions, and survival. In such countries, waste is collected in an uncontrolled way, and the products of their conversions migrate into the environment, posing health risks to humans. In many countries, including European, there is still a problem of "dumps", scattered mostly in forests around human settlements. Low environmental awareness and the lack of sense of responsibility for the environment could be discerned as the main reasons for this problem.

1.2 CURRENT STATE OF WASTE LANDFILLING

The approach to landfilling as a method of waste disposal has changed recently, although it is still popular in the world especially in developing countries. In Asian countries, open dumps and sanitary landfills are the main methods of solid waste disposal, with predominance these first. For example in Thailand and India, open dumps represents approximately 70–90% of the final disposal sites (Hogland et al., 2005). In developed countries the role of landfilling has significantly declined in recent years, opposite to the role of incineration, recycling or composting, that allow to material and/or energy recovery from waste. For example, in 2000 in the European Union (EU-27) approximately 55% of all waste generated, while in 2010 only 37% of them were deposited. The incineration share increased during this period from 15 to 21%, and material recycling from 14 to 24% (Eurostat 2005, Eurostat). High cost is a major barrier to the implementation of thermal methods of waste utilization. Biological treatment of waste in anaerobic conditions is becoming more and more popular, recently, and confirmed by the steadily growing number of installations using the anaerobic digestion for waste processing. According to Sidełko & Chmielińska-Bernacka (2013) in the years 1984–1994, 15 industrial installations for methane fermentation of waste were established in the EU. Over the next 10 years their number increased to 65. More than 70 plants treating biowaste or MSW were installed between 2006 and 2010. In 2012 in Europe there were 244 plants with a capacity to process almost 8 million tonnes of organic waste. Countries having the largest capacity installed were Germany (*ca.* 2 million tons) and Spain (1.6 million tons) (De Baere & Mattheeuws, 2012).

Despite the downward trend in waste landfilling in many countries around the world, the mass of waste going to landfills on the global scale is still significant. According to data by the World Bank, about 340 million tonnes of waste is placed in sanitary landfill and dumps each year, which accounts to roughly 40% of all waste produced in the world (Hoornweg & Bhada-Tata, 2012). According to data of Eurostat, in 2010, the production of municipal solid waste in the EU-27 was ca. 252.10 million tonnes. About 37% of this waste was deposited in landfills, equal to approximately 93.24 million tonnes (Eurostat). Assuming the average bulk density of compacted waste as 0.8 tonne/m³, the amount of waste deposited annually only in the EU-27 ranged up to 116.6 million m³. To illustrate the scale of the problem with this volume it is worth imagining a pyramid with a square base of side 1 km and a height of 350 meters (or 45 pyramids likes a Cheops ancient monument, with a side length of 230 m and a height of 147 m).

Generally, waste production rates are positively correlated to GDP, per capita energy consumption and final private consumption (Bogner et al., 2008). For example, India experienced an average GDP growth of 7% between 1997 and 2007, which is correlated with the increase in MSW of 48 million tonnes to 70 million tonnes during the same period (Chintan, 2009; Sharholy, 2008). Another important factor that influences waste generation is the level of urbanization. It is stated that urban residents produce about two times more waste than rural inhabitants (Hoornweg & Bhada-Tata, 2012).

Unfortunately, despite the significant progress that has already taken place in the field of waste management in many countries, the world still faces major challenges in this field. Taking under consideration the future trends in social and economic changes, it is estimated that the world MSW production is expected to increase from *ca* 1.3 billion tonnes (2010) to *ca.* 2.2 billion tonnes per year by 2025. This will correspond to an increase in per capita waste generation rates from 1.2 to 1.42 kg per person per day in considered period (Hoornweg & Bhada-Tata, 2012). A more substantial increase is expected in the rapidly developing countries, such as China or India, that do not have effective regulatory and policy instruments for waste minimization and recycling, as well as adequate facilities of waste pre-treatment and material/energy recovery.

Although municipal solid waste is not classified as a hazardous substance, the storage of it is unsafe for the environment due to the instability of its chemical composition. This is a result of a high content of organic matter susceptible to biodegradation. According to statistics from the World Bank, the content of organic fraction (mainly food and horticultural waste) in MSW composition ranged from at least 27% (in Europe) to 62% (in East Asia and Pacific countries) (Hoornweg & Bhada-Tata, 2012). The average value—estimated on the basis of global data—is 46% (Fig. 1.1). Thus, approximately 156 million tonnes of organic matter goes to landfills each year. A significant part of this matter undergoes biodegradation to mineral forms or simple organic substances that migrate to landfill gas and leachate or are retained in waste mass. Laboratory studies of refuse decomposition under controlled anaerobic conditions indicate that 25–40% of landfill carbon is converted to biogas carbon; however, under field conditions, the fraction of carbon converted would be less than in the laboratory (Bogner & Spokas, 1995).

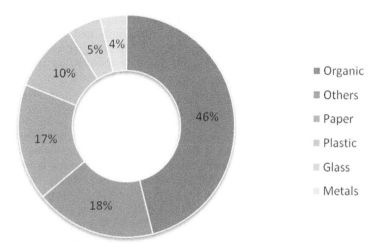

Figure 1.1   Global solid waste composition (Hoornweg & Bhata-Tata, 2012).

1.3   LANDFILL GAS IMPACT ON THE ENVIRONMENT

Waste deposited in landfills undergoes biological, chemical and physical transformations that cause changes in solid, liquid (leachate) and gas phases. As much as 90% of the carbon that is released during the decomposition of organic matter in the waste migrates to landfill gas in the form of CH_4, CO_2 and various hydrocarbons, whereas only 10% enters the leachate (Huber-Humer, 2007). The complex composition of the gas emitted from landfills and its uncontrolled emission into the atmosphere poses multiple threats and problems at different levels of coverage.

1.3.1   *Landfill gas contribution to climate change*

The emission of compounds—such as methane, carbon dioxide, nitrous oxide, and halogenated organic compounds, which have the ability to absorb infrared radiation, has the greatest extent on the landfill environmental impact, contributing to both global warming and climate change. Among the aforementioned gases, methane emissions from landfills are generally considered to represent the major source of climate impact in the waste sector. Landfill-derived methane represents 12% of all global methane emissions (EPA 2006). The significance of the gas in a greenhouse creation is assessed on the basis of its global warming potential (GWP) value. It is an index for estimating the relative global warming contribution due to atmospheric emission of 1 kg of a particular greenhouse gas compared to the emission of 1 kg of CO_2 (IPCC, 2001, I group report). The GWP value of methane is 25—when it is calculated in time horizon of 100 years, and 72—when a 20-year time horizon is applied (W&CC, 2010).

It should also be recognized that when landfill gas is captured and used to generate electricity, fugitive methane leaks from the landfill gas collection system, further contributing to total landfill GHG emissions (W&CC, 2010). The share of landfills in the global CO_2 emission is usually omitted. Lima et al., (2002) estimated that global biogenic CO_2 emissions from landfills is equivalent to 0.04–0.06% of fossil fuel emissions of CO_2. Landfills are also the source of nitrous oxide (N_2O) emission, where the atmospheric lifetime is estimated to be 114 years. The global warming potential of this gas is 298 in time horizon of 100 years (GWP100), and 289 in time horizon of 20 years (GWP20), as compared with CO_2 mass-to-mass (IPPC, 2007). Rinne et al. (2005) reported that N_2O emissions from one of the landfills in Finland amounted to 6.0 mg N m^{-2} h^{-1}, and it was about 1 order of magnitude higher than the highest emissions reported from Northern European agricultural soils, and 2 orders of magnitude higher than the highest emissions reported from boreal forest soils. Expressed as a greenhouse warming potential (GWP100), the N_2O emissions make up about 3% of the total GWP100 emission of the landfill in Finland.

1.3.2   *Landfill gas effect on the atmospheric chemistry*

Landfill gas burning (in a gas engine or flare) contributes to a significant reduction in GWP as a result of the oxidation of methane to CO_2, but also leads to the formation of nitrous oxide, that is produced by the oxidation of ammonia. This compound contributes to the creation of photochemical smog. Other nitrogen oxides; including NO and NO_2, that are also involved in a photochemical smog, are the products of landfill gas burning. Therefore, the ammonia concentration in landfill gas is often limited to 100 mg/m^3_n (Persson, 2006).

The non-methane organic compounds emitted from landfills contribute to photochemical smog creation. These compounds combine with nitrogen oxides present in the atmosphere to form the photochemical oxidants (e.g. ozone) (Cardelino & Chameides, 2000).

The reactions of methane with hydroxyl radicals (OH·)—very reactive and short-lived molecules—result in its oxidation, leading to a reduction of self-cleaning capacity of the atmosphere. Hydroxyl radicals, mainly arising from the reaction of photo dissociation of ozone in the presence of water vapour, fulfil the role of oxidants for almost all the gases getting into the atmosphere (Manahan, 1993). The concentration of OH·, and thus the concentration of several gases that react with them, depends primarily on the amount of CO and CH_4 in the troposphere, as almost 100% of the radicals reacts with these gases: 80% with CO, and 20% with CH_4 (Crutzen, 1994; Lu and Khalil, 1993). Hence, there is the claim that the increase in the CH_4 concentration reduces atmospheric self-cleaning capacity of other contaminants.

1.3.3   *Local odour nuisance*

On a local scale, the landfill impact on the environment is related to the emission of odorous gases. The gases that are mainly responsible for odour nuisance are: hydrogen sulphide, alkylated benzenes, esters, dimethyl sulphide (DMS), butyric acid, thiols (mercaptans), such as methanethiol and propanethiol (Parker et al. 2002), terpenes (p-cymene, limonene, α-pinen), carbon disulphide (Stretch et al. 2001) and ammonia. These gases are perceptible at very low concentrations in air, sometimes in order of a few ppm. Due to the odours nature and harmful effects of these gases on health, emissions of sulphur compounds from landfills should be limited. The presence of sulphur compounds in the biogas is also undesirable when the gas is used for energy production, because of its corrosive effects in the presence of water vapour. Particularly dangerous in this regard are oxidized sulphur compounds. They should be removed in order to avoid the destruction of compressors, gas storage tanks and piston rings in the engine.

1.3.4   *Human health hazards*

Many of the substances emitted from landfills may adversely affect living organisms. Some non-methane volatile organic compounds are cancerogenic. This group includes benzene, vinyl chloride, and tetrachloroethylene, classified by the International Agency for Research on Cancer (IARC) as Group-I carcinogens. This category is used for the agent (mixture) when there is sufficient evidence of their carcinogenicity on humans (IARC, 2006). Other compounds, such as tetrachloroethylene or trichloroethylene, are classified in Group 2A "probably carcinogenic to humans" and many compounds, such as bromodichloromethane, chloroform, 1,2-dichloroethane, dichloromethane, ethyl benzene and styrene are classified to Group 2B "possibly carcinogenic to humans". Toxic substances are also generated during landfill gas combustion. When halogenated chemicals (chemicals containing halogens—typically chlorine, fluorine, or bromine) are combusted in the presence of hydrocarbons, they can recombine into highly toxic compounds such as dioxins and furans, the most toxic chemicals ever studied (EPA, 1994).

Dust particulates which have migrated to the environment from the landfill gas also pose a danger to human health. The distance of their relocation depends on particle size and wind speed. According to data from the UK Department of Environment (1995), cited in a document by the Health Protection Agency, titled "Impact on Health of Emissions from Landfill Sites" (2011) the coarse particulates (>30 µm) from quarries are deposited within 100 m from the source, while fine particles (<10 µm) can travel up to 1 km.

Landfills pose additionally a microbiological risk for the environment because they are the source of bioaerosols. Bioaerosols are defined as aerosols or particulate matter of microbial, plant or animal origin that are often used synonymously with organic dust. Bioaerosols consist of pathogenic or non-pathogenic live or dead bacteria and fungi, viruses, high molecular weight allergens, bacterial endotoxins, mycotoxins, pollen, etc. (Douwes et al. 2003). The total number of bacteria cells in the air on the landfill under the operation phase and in the waste sorting station ranged from 10^2 to 10^4 CFU m^{-3} (Breza-Boruta, 2012, Burkowska et al. 2011). High concentration of actinomycetes, *Pseudomonas flourescens*, potentially pathogenic bacteria belonging to the family *Enterobacteriaceae*, such as *Salmonella* sp. and *Escherichia coli*, faecal streptococci (Breza-Boruta, 2012), haemolytic bacteria, microscopic fungi belonging to *Penicillium, Cladosporium, Aspergillus, Alternaria, Fusarium, Acremonium, Drechslera*, and yeasts, mainly of genus *Candida* were found (Burkowska et al. 2011). It was also stated that concentrations of microorganism cells in air over the landfill were higher in winter than in other seasons (Huang et al. 2002).

The risk of explosion or fires from landfill gas are another category of human health risks or life threats. Landfill gas becomes explosive when it escapes from the landfill and mixes with oxygen. The lower explosive limit is 5% methane and the upper explosive limit is 15% methane. There are many reports regarding the gas explosions at landfills which had serious consequences. In 1986 in Loscoe (Derbyshire, UK) the explosion of landfill methane destroyed a bungalow on a housing estate near a landfill site, badly injuring its owners and causing post-traumatic stress for residents of the locality (May, 2011). Another tragic incident took place in Winston-Salem (South Carolina, USA) in 1969. Three people died and a few others were severely burned after the explosion of the landfill gas which had accumulated in the one of room in the National Guard Armory located near the landfill (City of Winston-Salem Directing Board, 1960–1960). Furthermore, there are many reports on fires breaking out at landfills in various parts of the world (Ettala et al. 1996, U.S. Fire Administration 2001, Landfill fires 2002, Vassiliadou et al. 2009).

Migration of landfill gas into the surrounding soils can also result in the deterioration of plant growth. The lack of oxygen, caused by its displacement of landfill gas, contributes to the damaging the root system in plants.

1.4   ROLE OF WASTE LANDFILLING IN CARBON BUDGET

In the context of numerous threats, the question arises, can we find any positive aspects of waste landfilling? Recently there are more and more voices pointing to the positive aspects of the waste deposition. They emphasize that when the appropriate level of technical security is assured, landfills must not be seen only as a source of environmental problems. It should be borne in mind that they may also play the positive role in the environment.

1.4.1   *Landfills as a carbon repository*

Recently, the question regarding the carbon sequestration in order to prevent climate change was raised. Global-scale observations of the chemical composition of the atmosphere show that we must be prepared to huge imbalances between the emission of carbon into the atmosphere, and its assimilation by all possible sinks on the ground. Thus, carbon storage outside the atmosphere, in a stable form as long as possible, is highly advisable. The combustion of organic waste does not fit this scenario if the produced CO_2 is not captured and stored. Only small amounts of carbon (less than <1%) remains in the ashes (IPCC, 2007). Storage of carbon-reach waste may be considered a long-term carbon deposition method. The analysis shows that over 50% of the carbon contained in the waste accumulated in landfills is not transferred to gas or leachate but remains trapped in the mass of waste for a long time. This is a larger amount of carbon than that retained in composted waste, in which 15–50% of carbon is not released to gas or liquid phase (IPCC, 2007). Lower susceptibility of organic compounds to biodegradation in anaerobic conditions promotes greater carbon trapping. According to Barlaz (2006) some organic materials deposited in landfills, such as plastics, are non-biodegradable, while others, such as lignin, lignin-bound cellulose and hemi-cellulose, undergo minimal decomposition in the anaerobic conditions. Thus, deposition of wood waste under anaerobic conditions in landfills may be considered as carbon storage influencing the GHG emissions. Moreover, some portion of organic matter is subjected to humification, forming humic acids resistant to biodegradation. Zach (2002) stated that the content of humic acids in a 15 year old landfill was 22% of VS. Carbon that is immobilized in landfills does not release to the natural geo-ecosystem. Manfredi et al. (2009) taking into account over a 100-year time horizon, calculated that carbon stored in European sanitary landfills will lead to GHG savings of 132 to 185 kg CO_2-eq per tonne of wet mixed MSW. In this context, waste landfilling can be seen as a way of carbon sequestration (Cossu et al. 2007).

1.4.2   *Landfill gas as renewable energy source*

Landfill gas is biogenic in origin and therefore its combustion for energy production instead of fossil-fuels can reduce GHG emissions. The benefits of waste landfilling for the global climate are largely dependent on the type of fossil fuel that is assumed to be replaced by LFG (W&CC, 2010). Monni et al. (2006) calculated that according to the prediction for 2030 replacement of natural gas in electricity generation by LFG will lead to GHG emission savings from 16 to 49 Tg CO_2eq depending on the type of waste management scenario. High emission savings, ranging from 42 to 126 Tg CO_2eq, can be obtained when coal is replaced by LFG.

1.5   STRATEGIES OF MITIGATION FOR LANDFILL GAS EMISSION

Current available technologies allow for significant minimization of many negative environmental effects caused by landfills. Activities that allow this goal to be achieved can be realized at different levels, such as:

- control of the waste input; minimization of the waste mass that is sent to the landfill: separation of hazardous components from MSW; recycling, material and energy recovery,
- control of the landfill reactor operation by physical barriers; bottom liners and capping systems, temporary and daily covers, gas drainage and collection system: by chemical methods—liming, aeration, leachate recirculation,
- control of gas release to the environment: gas burning (for energy recovery, in flare), and biofiltration in porous materials.

The main strategies of LFG emission mitigation is based on three milestones:

1. Abatement of the gas amount due to a decrease in the total LFG production. This strategy is based on reducing the total amount of waste deposited in landfills, but particularly on limitating biodegradable waste storage.
2. Shortening the time of the gas emissions by accelerating the process of anaerobic digestion under controlled conditions. It can be obtained by creating optimal conditions for the

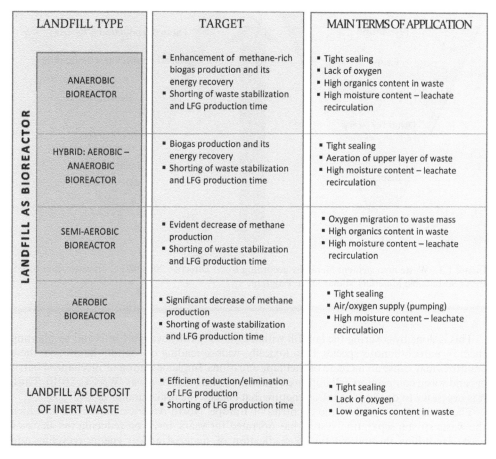

LANDFILL TYPE		TARGET	MAIN TERMS OF APPLICATION
LANDFILL AS BIOREACTOR	ANAEROBIC BIOREACTOR	• Enhancement of methane-rich biogas production and its energy recovery • Shorting of waste stabilization and LFG production time	• Tight sealing • Lack of oxygen • High organics content in waste • High moisture content – leachate recirculation
	HYBRID: AEROBIC – ANAEROBIC BIOREACTOR	• Biogas production and its energy recovery • Shorting of waste stabilization and LFG production time	• Tight sealing • Aeration of upper layer of waste • High moisture content – leachate recirculation
	SEMI-AEROBIC BIOREACTOR	• Evident decrease of methane production • Shorting of waste stabilization and LFG production time	• Oxygen migration to waste mass • High organics content in waste • High moisture content – leachate recirculation
	AEROBIC BIOREACTOR	• Significant decrease of methane production • Shorting of waste stabilization and LFG production time	• Tight sealing • Air/oxygen supply (pumping) • High moisture content – leachate recirculation
LANDFILL AS DEPOSIT OF INERT WASTE		• Efficient reduction/elimination of LFG production • Shorting of LFG production time	• Tight sealing • Lack of oxygen • Low organics content in waste

Figure 1.2   Approaches to minimize the landfill impact on the atmosphere.

growth of microorganisms, which enhances the production of biogas suitable for energy generation. Although the biogas production is high in this case, its uncontrolled release into the environment is minimized by appropriate barriers. The phase in which the lean LFG is produced is shortened. Wetting the waste body by leachate recirculation improves the transport of nutrients and micro-organisms that results in an increase of decomposition rate and in depletion of substrates susceptible to biodegradation.

3. Diversion of a direction of biochemical transformation taking place in the waste mass from anaerobic to aerobic leads to minimization of methane release in favour of carbon dioxide with lower GWP.

Different ways leading to minimise the landfill impact on the atmosphere and terms of their applications are presented in Figure 1.2.

In practice, the choice of the strategy is a part of the environmental policy and the derivative of the economic status of the countries in which they are implemented. In most countries in Asia and Africa, a priority in the field of waste management is given to the separation of the municipal waste stream of hazardous waste and the development of methods for hazardous waste facility. With regard to waste deposition, the main task is to change the storage system from open dumps to sanitary landfill, and minimize the environmental impact of existing open dumps.

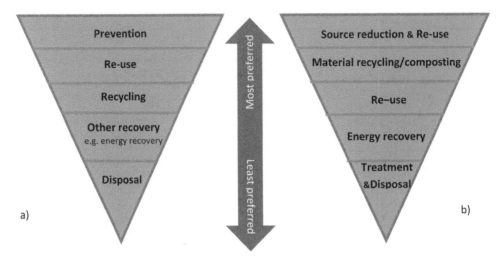

Figure 1.3   Waste management hierarchy according to: a) Directive 2008/98/EC on waste, and b) US EPA's Sustainable Materials Management Program.

This is done by covering the landfill with a layer of locally available soils and by planting them by autochthonous species. Paradoxically, wide-spreading of sanitary landfills in these countries may cause an increase of methane emissions. Implementation of mechanical barriers and waste compaction will improve conditions for methanogenesis (W&CC, 2010). Thus, it is necessary to develop the infrastructure that enables the utilization of landfill gas.

The strategies adopted by the countries of Europe, North America, and Japan, in which the waste stream separation system has operated for years, insist on reducing the share of waste landfilling throughout the intensification of material and/or energy recycling and combustion. As a result, the landfill sites should be used as a place where chemically stable materials, resistant to biochemical transformation, will be stored. It is aimed at eliminating the long-term impact of landfills on the environment. In EU countries, such operations arise from the adopted environmental policies regarding waste management presented in the Waste Framework Directive (Directive 2008/98/EC). The Directive lays down fundamental principles of waste management. It requires that waste should be managed in such a way that does not damage human health or the environment, does result in an odour nuisance, and does not adversely change the countryside or places of special interest. Priority is given to minimise waste production, re-use, recycling, waste-to-energy recovery, and finally disposal. Landfilling excludes a certain area of the earth's surface from natural use for a long time, changing it into the source of emissions of environmentally hazardous substances that pose a threat to groundwater thorough, soils, and especially air. Therefore, landfilling is placed at the bottom of the pyramid used to visualise the waste management hierarchy (Fig. 1.3a). It is the least desirable option and should be gradually reduced year by year. A similar hierarchy of waste management is in force in the U.S. (Fig. 1.3b).

All the treatments that preceded waste landfilling should reduce the quantity of municipal solid waste that is sent to landfills. The aforementioned Directive obligates EU members to minimize the amount of biodegradable municipal waste to landfill to the level corresponding to:

• 75% in 2006 (2010 for the countries that have been granted a 4 year derogation)
• 50% in 2009 (2013, as above)
• 35% in 2016 (2020, as above)

referencing to the amount of waste deposited in 1995 (100% by weight).

The decision regarding the selection of the method of waste treatment, limiting the negative impact of landfills, is affected by many factors. The most important include:

– landfill status (existing, proposed),
– landfill type (open dump, sanitary landfill),
– age of the landfill, influencing the amount and composition of LFG,
– size of the landfill, influencing the amount of gas,
– geological and climatic conditions, such as type of soil, ground water level, rainfall,
– financial capacity of the investor,
– principles of environmental law.

There are many roads that lead to the reduction of greenhouse gas emission from landfills, yet none of them are self-sufficient. The most efficient results can be achieved when the actions are undertaken on different levels by using different methods. For large landfills, with a high organic matter content, the most effective solution may be, according to Huber-Humer (2004), the combination of gas utilization for energy production with landfill bio-cover specially designed for efficient CH_4 oxidation. The combustion of biogas, despite the clear benefits (economic and environmental), encounters many obstacles such as high cost of installing a gas recovery and burning systems. It is a main brake in their dissemination in developing countries.

Emissions from solid waste disposal are expected to increase with increasing global population and GDP. Global emissions from landfills are projected to increase from 340 Tg CO_2eq in 1990 to 1500 Tg CO_2eq by 2030 and 2900 Tg CO_2eq by 2050 in the Baseline scenario in which waste generation is assumed to follow past and current trends using population and GDP as drivers. The total mitigation potential are calculated to be up to 30% in 2030 and 50% in 2050 assuming dissemination of incineration, recycling and landfill gas recovery (Monni et al. 2006).

It is important in the public interest to develop and implement the consistent standards of waste landfilling practice that will allow to maximally reduce the impact on the environment. Developing today the idea of "sustainable landfill" emphasizes this aspect of storage. Cossu (2007) specifies a sustainable landfill as a system that should reach an acceptable equilibrium with the environment within one generation (30–40 years). This phrase links to the base definition of "sustainable development" given by the Brundtland Commission in its 1987 report Our Common Future: "development that meets the needs of the present without compromising the ability of future generations to meet their own needs (...)." Each generation should manage the waste in such a way that does not leave the problem for the next generation, which has the right to live in a clean and safe environment. But the roads that lead to reaching this state of the landfill, which can be called sustainable, are not strictly defined. According Huber-Humer et al. (2010) a "sustainable landfill is understood as a landfill where the disposed of waste mass is already in a stable state (...) and emission release is below the local environmentally acceptable level or can be controlled by simple, robust and natural measures, e.g. control of gas emissions by methane oxidation in landfill covers". Additionally such a landfill can be regarded as a long-term carbon and nitrogen storage pool. Kavazanjian draws attention to maximizing the capacity of existing landfills and minimizing the need for new landfills, and making beneficial use of closed landfill sites. Appropriate re-use of the landfill allows for equilibrium with the environment, restoring the area for the nature or giving it a new practical function.

1.6 SUMMARY

Taking into account the expected directions of economic and social changes around the world, the increase of municipal solid waste production is predicted. At the current level of technological development, it is possible to obtain a very efficient reduction of the waste management sector impact on the environment. However, the problem with applying modern solutions, lies mainly in the high costs, which for the majority of developing countries are an insurmountable barrier. Efforts to reduce the amount of landfilled waste require a large financial contribution to the development of segregation and recycling technologies. Therefore, it is expected that many countries will keep landfilling as the primary method of waste disposal. Special attention should be paid to waste landfilling conditions. All the open dumps should be replaced by sanitary landfills, which provide opportunities to control emissions and migration of leachate into the environment (IPCC, 2007). According to Manfredi et al. (2009) and Bahor et al. (2009) looking forwards to next 100-year period, properly managed landfills of the type seen in developed countries, may capture around 50–80% of methane generated. To make this scenario come true, it is necessary to elaborate the relevant regulations and mechanisms for its respect. Assistance from countries that have the knowledge and experience, and are also able to provide financial support, will be greatly needed.

REFERENCES

Bahor, B., Van Brunt, M., Stovall J., Blue, K. 2009. Integrated waste management as a climate change stabilization wedge. *Waste Management and Research* 27: 839–849.

Barlaz, M.A. 2006. Forest products decomposition in municipal solid waste landfills. *Waste Management* 26(4): 321–333.

Bogner, J., Spokas, K. 1995. Carbon storage in landfills, Chapter 5. In: Lal, R. et al. (eds). *Soils and global change. Advances in Soil Science Series.* CRC Lewis Publications, Boca Raton, Florida.

Bogner, J., Pipatti, R., Hashimoto, S., Diaz, C., Mareckova, K., Diaz, L., et al. 2008. Mitigation of global greenhouse gas emissions from waste: conclusions and strategies from the Intergovernmental Panel on Climate Change (IPCC) fourth assessment report. *Waste Management and Research* 26: 11–32.

Breza-Boruta, B., 2012. Bioaerosols of the municipal waste landfill site as a source of microbiological air pollution and health hazard. *Ecological Chemistry and Engineering* 19 (8): 851–862.

Burkowska, A., Swiontek Brzezinska, M., Kalwasińska, A. 2011. Impact of the municipal landfill site on microbiological contamination of air. *Contemporary Problems of Management and Environmental Protection,* No. 9. Some Aspects of Environmental Impact of Waste Dumps: 71–87.

Cardelino, C.A. & Chameides, W.L. 2000. The application of data from photochemical assessment monitoring stations to the observation-based model. *Atmos. Environ.* 34: 2325–2332.

Chintan 2009. *Cooling agents: an analysis of climate change mitigation by the informal recycling sector in India.* Report prepared in association with The Advocacy Project, Washington DC.

City of Winston-Salem Directing Board: 1960–1969. *City of Winston-Salem Government Meetings Notes* (http://www.cityofws.org/portals/0/pdf/marketing-and-communications/WinstonSalem%2019601969.pdf).

Cossu, R., 2007. Landfill sustainability—concepts and models, in: *Proceedings Sardinia 2007, Eleventh International Waste Management and Landfill Symposium,* 1–5 Oct. 2007, S. Margherita di Pula-Cagliari, Sardinia, Italy, CISA 2007. Bogner, J.; Spokas, K., Carbon storage in landfills, Edited by Lal, Rattan, Soils Global Change: 67–80, 1995.

Crutzen, P.J. 1994. Global Budgets for Non-CO$_2$ Greenhouse Gases. *Monitoring and Assessment* 31, 1–15.

De Baere, L. Mattheeuws, B. 2012. Anaerobic Digestion of the Organic Fraction of Municipal Solid Waste in Europe—Status, Experience and Prospects. *Waste Management* 3: Recycling and Recovery—Thomé-Kozmiensky Karl J., Thiel S., TK-Verl., Nietwerder: 517–526.

Directive 2008/98/EC of the European Parliament and of the Council of 19 November 2008 *on waste and repealing certain Directives.* Official Journal of the European Union. L 312/3.

Douwes, J., Thorne, P., Pearce, N., Heederik, D. 2003. Bioaerosol health effects and exposure assessment: Progress and prospects. *The Annals of Occupational Hygiene* 47(3): 187–200.

EPA 1994. *Dioxin reassessment, estimating exposure to dioxin-like compounds,* Vol. 2, Chapter 3 (http://www.cqs.com/epa/exposure).

EPA 2006. *Solid waste management and greenhouse gases—A life-cycle assessment of emission and sinks.* 3rd Edition. U.S. Environmental Protection Agency. Washington, DC. September 2006.

EPA SMM. *US EPA's Sustainable Materials Management Program* (http://www.epa.gov/smm).

Ettala, M., Rahkonen, P., Rossi, E., Mangs J., Keski-Rahkonen, O. 1996. Landfill Fires in Finland, *Waste Manag Res*. 14(4): 377–384.

Eurostat website (http://epp.eurostat.ec.europa.eu/portal/page/portal/waste/introduction).

Eurostat 2005. *Waste generated and treated in Europe*. Data 1995–2003. European Commission-Eurostat, Luxemburg.

Health Protection Agency, 2011. *Impact on Health of Emissions from Landfill Sites*, RCE-18 (http://www.hpa.org.uk/webc/hpawebfile/hpaweb_c/1309969974126).

Hogland, W., Visvanathan, C., Marques, M. and Manandhar, D.R. 2005. Landfill in Asia Improving sanitation of landfill sites. *Waste Management World*, July–August 2005: 87–96.

Hoornweg, D. Bhada-Tata, P. 2012. What a waste: a global review of solid waste management. Urban development series: knowledge papers no. 15. Washington, DC: World Bank. (http://documents.worldbank.org/curated/en/2012/03/16537275/waste-global-review-solid-waste-management).

Huang, C.-Y., Lee C.-C., Li, F.-C., Ma, Y.-P., Su, H.-J.J. 2002. The seasonal distribution of bioaerosols in municipal landfill sites: a 3-yr study. *Atmospheric Environment* 36(27): 4385–4395.

Huber-Humer, M. 2004: *Abatement of landfill methane emissions by microbial oxidation in biocovers made of compost*. PhD Thesis, University of Natural Resources and Applied Life Sciences (BOKU), Vienna.

Huber-Humer, 2007. Dwindling Landfill Gas-Relevance and Aftercare Approaches. In P. Lechner (ed). *Waste matters. Integrating views* 2nd BOKU Waste Conference. Facultas Veralgs- und Buchhandels AG.

Huber-Humer, M, Smidt, E., Tinter, J., Gamperling, O., Böhm, K., Lechner, P. 2010. New concept and methods to evaluate the sustainability of landfills ISWA World Congress, November 2010, Hamburg, Germany.

IARC, 2006. *Monographs on the Evaluation of Carcinogenic Risks to Humans*. Preamble. World Health Organization International Agency for Research on Cancer. Lyon, France 2006.

IPCC, 2001. *Climate Change: The Scientific Basis*. Contribution of Working Group I to the Third Assessment Report of the Intergovernmental Panel on Climate Change, Working Group I. J.T. Houghton, Y. Ding, D.J. Griggs, M. Noguer, P.J. van der Linden, X. Dai, K. Maskell, C.A. Johnson (eds). Cambridge University Press.

IPCC, 2007. *Climate Change: Mitigation*. Contribution of Working Group III to the Fourth Assessment Report of the Intergovernmental Panel on Climate Change. Chapter 10-Waste management. J. Bogner, Coordinating Lead Author. B. Metz, O.R. Davidson, P.R. Bosch, R. Dave, L.A. Meyer (eds). Cambridge University Press.

Kavazanjian, E. 2006. Sustainable Landfilling. NRC Board of Earth Sciences Roundtable. (http://dels.nas.edu/besr/docs/04-Kavazanjian.pdf).

Landfill fires their magnitude, characteristics, and mitigation, May 2002/FA–225 (http://www.usfa.fema.gov/downloads/pdf/publications/fa-225.pdf)

Lima, R., Nolasco, D., Meneses, W., Salazar, J., Hernández, P., Pérez, N. 2002. *Global Biogenic Emission of Carbon Dioxide from Landfills*. American Geophysical Union, Fall Meeting 2002.

Lu, Y., Khalil, M.A.K. 1993. Methane and Carbon Monoxide in OH Chemistry: the Effects of Feedbacks and Reservoirs Generated by the Reactive Products. Chemosphere 26 (1–4): 641–655.

Manahan, S.E. 1993. *Fundamentals of environmental chemistry*. Lewis Publishers (Boca Raton Fla.).

Manfredi, S., Scharff, H.M., Tonini, D., Christensen, T.H. 2009. Landfilling of waste: accounting of greenhouse gases and global warming contributions, *Waste Manag Res*. 27 (8): 825–836.

May, C. 2011. *Laws Tightens for Landfill Operators Waste Management World Waste Management World Waste Management World* 12(2). (http://www.waste-management-world.com/articles/print/volume-12/issue-2/features/laws-tightens-for-landfill-operators.html).

Monni, S., Pipatti, R., Lehtilä, A., Savolainen, I., Syri, S. 2006. *Global climate change mitigation scenarios for solid waste management*. VTT Publications 603. VTT Technical Research Centre of Finland, Espoo, Finland.

Our Common Future. 1987. *Report of the World Commission on Environment and Development*. United Nations, (http://conspect.nl/pdf/Our_Common_Future-Brundtland_Report_1987.pdf).

Parker, T.J., Dottridge, S. Kelly, Investigation of the composition and emissions of trace components in landfill gas. R&D Technical Report P1-438/TR. Environment Agency 2002.

Persson, M., Jönsson, O., Wellinger, A. 2006. *Biogas upgrading to vehicle fuel standards and grid injection*, IEA Bioenergy.

Rinne, J., Pihlatie, M., Lohila, A., Thum, T., Aurela, M., Tuovinen, J.P., Laurila, T. Vesala, T. 2005. Nitrous oxide emissions from a municipal landfill. *Environ Sci Technol*. 39(20): 7790–3.

Sharholy, M., Ahmad, K., Mahmood, G., Trivedi, M.C. 2008. Municipal solid waste management in Indian cities-A review. *Waste Manag* 28(2): 459–67.

Sidełko R., Chmielińska-Bernacka A. 2013. Application of compact reactor for methane fermentation of municipal waste. *Anual Set The Environment Protection* 15: 683–693.

Stretch, D., Laister, G., Strachan, L., Saner, M. 2001. Odour trails from landfill sites. In *Proceedings of 8th International Waste Management and Landfill Symposium*; Sardinia 2001. Cagliari, Italy: 709–718.

U.S. Fire Administration, 2001. Topical Fire Research Series 1(18), (http://www.usfa.dhs.gov/downloads/pdf/tfrs/v1i18–508.pdf).

Vassiliadou, I., Papadopoulos, A., Costopoulou, D., Vasiliadou, S., Christoforou, S., Leondiadis, L. 2009. Dioxin contamination after an accidental fire in the municipal landfill of Tagarades. Thessaloniki, Greece. *Chemosphere* 74: 879.

W&CC, 2010. *Waste and Climate Change: Global trends and strategy framework*. United Nations Environmental Programme Division of Technology, Industry and Economics International Environmental Technology Centre Osaka/Shiga.

Zach, A. 2002. Characterisation of organic substances in stabilized composts of rest wastes. In: H. Insam, N. Riddech, S. Klammer (eds). *Microbiology of Composting*. Springer Berlin Heidelberg.

CHAPTER 2

Characteristics of landfill gas

2.1 INTRODUCTION

Municipal waste deposited in non-hazardous landfills usually contains about 50% of organic matter, which is generally susceptible to microbial degradation. Commonly applied waste compaction which aims at the reduction of their volume, as well as the use of barriers insulating the mass of waste from the environment favor the creation of anaerobic conditions. In such conditions, consortia of fermenting microorganisms are developing. They use organic compounds or intermediate products of their decomposition, contained in waste, as electron acceptors in the process of energy production. This group includes the microorganisms that produce methane. A large group are also sulfur bacteria, sulfate-reducing bacteria and bacteria leading to nitrogen transformations (ammonification and denitrification). As a result of these changes, gaseous products or water-soluble compounds are formed. Gases and volatile liquids easily migrate to the environment. According to the Article 2 of Council Directive 1999/31/EC on the landfill of waste "all the gases generated from the landfilled waste" are called "landfill gas". This gas is produced mainly by microbiological processes. However, also abiotic processes such as chemical reactions between components of waste and physical phenomena, such as evaporation or sublimation, also have little influence on its composition; especially with regard to trace gases (EPA, 2008).

2.2 LANDFILL GAS COMPOSITION

2.2.1 *Factors influencing landfill gas composition*

Taking into consideration the mechanism of formation, landfill gas (LFG) is a type of biogas, similar to mixture of gases generated inside sewage sludge digesters and farm plant. However, in comparison to the gas formed in installations, used to methane fermentation, to which the substrate is transported continuously, LFG is characterized by greater variability of composition (Table 2.1). It results mainly from a separation of particular phases of biochemical changes in the landfill, a greater variety of morphological fermented mass; as well as exposure to the influence of external factors such as pressure, temperature, and atmospheric precipitation. The landfill can be compared to the anaerobic batch reactor, where organic matter, introduced once, is gradually decaying, until the exhaustion of all material which is susceptible to biodegradation. In the case of landfill, the waste deposition phase is stretched during the whole exploitation phase and usually takes several years, whereas the whole decomposition process takes many decades. The decomposition of organic matter in landfill consists of several consecutive phases. Reinhart & Townsend (1998) distinguish initial phase, transitional phase, acid production phase, methanogenic phase and maturation phase. Kjeldsen et al. (2002), referring to earlier works (Christensen & Kjeldsen, 1995; Bozkurt et al. 2000) distinguish initial aerobic phase, anaerobic acid phase, initial methanogenic phase, stable methanogenic phase and aerobic or humic phase. Products of each preceding phase are substrates for the next one. They dissolve in water contained in waste or are released to the atmosphere as a gasses. In consecutive phases, together with degradation and stabilization of organic wastes, the gas composition changes and chemical compounds, typical for particular phases, occur in leachate. Initial aerobic phase begins immediately after the deposit of waste on the landfill when oxygen is present in free spaces between waste and pores of solid

Table 2.1 Composition of landfill gas and biogas from farm plants and sewage sludge anaerobic digesters.

Object	Gas components								References
	CH$_4$ [vol.%]	CO$_2$ [vol.%]	O$_2$ [%]	N$_2$ [%]	H$_2$S [ppm]	NH$_3$ [%]	H$_2$ [%]	CO [%]	
Landfill	45–60	40–60	0.1–1	2–5	0–1000*	0.1–1	0.02	0–0.2	Tchobanoglous et al. 1993
Landfill (values from 7 waste disposal sites in UK)	37–62	24–29	<1	<1–2	n.a	n.a	n.a	n.a	Allen et al. 1997
Landfill, Korea	59.4–67.9	29.9–38.8	n.a.	n.a	15.1–427.5	n.a	n.a	n.a	Shin et al. 2002
Fresh Kills Landfill, USA	55.63	37.14	0.99	n.a	n.a	n.a	n.a	n.a	EPA, 1995
Grand' Landes landfill, France	37	25	7	32	n.a	n.a	n.a	n.a	Scheutz et al. 2008
Landfill	44	40.1	2.6	13.2	240	n.a	n.a	n.a	Jaffrin et al. 2003
Landfill – LFG from collector pipe	47–57	37–41	<1	<1–17	36–115	n.a	n.a	n.a	Rasi et al. 2007
Sewage sludge digester (MSTP)	61–65	36–38	<1	<2	b.d	n.a	n.a	n.a	
Farm biogas plant (cow manure & confectionary by-products)	55–58	37–38	<1	<1–2	32–169	n.a	n.a	n.a	
Sewage sludge digester	57.8	38.6	0	3.7	62.9	n.a	n.a	n.a	Spiegel et al. 1999
Landfill during operation						n.a	n.a	n.a	Dudek et al. 2007
– LFG from extraction wells	47.2–6.8	31.2–35.4	0–3	n.a	n.a				
– LFG from borehole	61.2	23.6	0						
Landfill during operation – waste without compaction	8.0–26	15.0–24.1	0–1	n.a	n.a	n.a	n.a	n.a	Meres et al. 2004
Landfill, Poland						n.a	n.a	n.a	
– part during operation	59–63	n.a	n.a	n.a	n.a				
– reclaimed part	42–62								
Landfill, Poland						n.a	n.a	n.a	Nowakowski, 1997
– gas stream freely releasing extraction well	60–65	34–38	0	n.a	133*				
– gas sucked from the landfill adequately isolated	45–50	35–45	1–4		100*				
– gas sucked from the landfill inadequately isolated	25–45	20–35	4–10		100*				

bodies. This phase begins with the hydrolysis of macromolecular substances, and then bio-degradation of the hydrolysis products with the participation of oxygen heterotrophs takes place. After the exhaustion of oxygen, in the anaerobic conditions, phases that are typical for methane appear one after another, namely hydrolysis of macromolecular substances, acid fermentation, transformation of organic acids into acetate (methanogenic phase also called unstable octanogenic) and methane production (methanogenic stable phase). The last, fifth phase (maturation or oxidation) occurs when biodegradable organic compounds are exhausted in landfill deposits.

Many years of observation showed that phases that take place in the mass of deposited waste are related with the age of landfill, which is understood as the time elapsed since the deposition of the waste or the appearance of leachate (Andreottola & Cannas, 1992). Due to the age, the landfills are divided into: young—operated for less than five years, the medium—from 5 to 10 years and old (mature)—over 10 years (Chian & De Walle, 1976). The young landfills are in a phase prior to the methanogenic one or in the early unstable methanogenesis. The medium landfills are characterized by stable methanogenesis, whereas the process of maturation begins on the old landfills.

The composition of the gas mixture emitted from the layer of waste changes with the age of the landfill. In the initial stages of transition in the biogas, nitrogen and oxygen, contained in the pores of the waste, prevail. There are also carbon dioxide and odor-generating gases such as methanethiol (methyl mercaptan) and dimethyl sulfide, which are biochemical products. When the oxygen is exhausted the fermentation process begins. The products of fermentation are mainly low molecular weight organic acids and alcohol, carbon dioxide and hydrogen. Over time, the concentration of carbon dioxide and hydrogen is increasing, reaching a maximum value of 80 vol.% and 20 vol.%, respectively, in octanogenic phase. This phase is characterized by intense secretion of compounds of odor character, namely fatty acids, alcohols and aldehydes (Parker et al. 2002). Methane appears in the biogas only when the methanogenic bacteria begin to metabolize the intermediate products of decomposition, such as acetate, hydrogenocarbonates, hydrogen, and carbon dioxide. When the number of methanogens reaches an appropriate level, the dynamic equilibrium is established between the rate of production of organic acids and acetate and methane, so that the concentration of methane in the biogas is stabilizing. As a result of intensive use of acetates and the operation of autotrophic methanogens using carbon dioxide and hydrogen, the methane content in the biogas in the methanogenic stable phase exceeds the concentration of carbon dioxide. The ratio of CH_4 to CO_2 concentrations in this phase may exceed the value of 1.6 (Lebedev et al. 1994). During methanogenic stable phase the total content of CH_4 and CO_2 concentration constitute 98–99% of landfill gas volume (Soltani-Ahmadi, 2000). Also odorous compounds (such as hydrogen sulphide, mercaptans, ammonia) and nitrogen are present in the biogas generated in this phase; whereas hydrogen, which is used by autotrophic methanogenic bacteria, it disappears. Over time, the supply of organic matter contained in the waste is exhausted. It leads to the decline in biogas production and methane reduction below the CO_2 concentration. Due to the reduction of gas pressure within the landfill, atmospheric air begins to diffuse to waste mass. When the rate of diffusion of oxygen is higher than the rate of its consumption by the microorganisms, aerobic conditions may occur in the interior of the landfill (Kjeldsen et al. 2002). Therefore, in this phase; microbial oxidation of CH_4, and other trace compounds that migrate from deeper layers of landfill, may take place. Changes during this phase are not exactly recognized as research in this area were carried out for too short a period. Available results refer to the past four decades. Although some authors have argued that due to the low gas emission, the landfills in the second oxygen phase constitute a low risk, the lack of experimental data hinders unequivocal acceptation or rejection of this hypothesis.

Apart from the phase nature of landfill transformations the landfill gas composition is also affected by the operating conditions and meteorological parameters. The landfill gas extraction system also plays very important role here. According to Nowakowski (1997), the differences in methane concentration between the gas stream flowing freely from the degassing well and the gas that is actively sucked reach even tens of percent. The lowest concentration of methane was found in the gas sucked from the landfill inadequately sealed. Negative pressure causes the suction of ambient air, which leads to an increase in the oxygen concentration in biogas (Table 2.1) and reduction in the concentration of halogenated aromatic hydrocarbons, from 5–100 mg m^{-3} in the gas flow freely escaping from extraction well, to the values below 50 mg m^{-3} in the gas sucked from the landfill (Nowakowski, 1997). According to Meres et al. (2004), methane concentration in LFG at the gas collection network was more stable when the pressure in the installation was above the ambient atmospheric value. When the pressure inside the network was less than atmospheric, significant fluctuations of methane content, of up to 15% vol., were measured. They also stated that high precipitation, lowering the permeability of the outer layer of the landfill and low air temperature improved the quality of biogas in the gas collection system. Zou et al. (2003) observed that the concentration of halogenated compounds and alcohols is decreasing over time, whereas the concentration of aromatic compounds is increasing.

Moreover, they found a greater variety and higher concentrations of NMVOCs in the landfill environment during the summer (60 compounds) than in winter (38 compounds). They explained it by faster biodegradation of organic matter in the waste, at higher temperatures and air humidity. In the summer, the concentrations of the most analyzed compounds were from several to ten times higher than those measured in winter.

2.2.2   *Characteristics of landfill gas components*

Gas generated inside a landfill, like other biogases consists mainly of methane (CH_4) and carbon dioxide (CO_2). Their concentration in methanogenic stage varied from 30 to 70 vol.% for CH_4 and from 20 to 50 vol.% for CO_2 (Table 2.1). Both the gases are odorless and colorless, but only methane is combustible. Lower heating value of methane is about 35 MJ m$_n^{-3}$. Both gases absorb infrared radiation, thus they are responsible for the greenhouse effect.

Landfill gas contains also a small amount of oxygen (O_2) and nitrogen N_2. Total concentration of these gases is usually below few percent (Table 2.1). The highest concentration of them are observed under aerobic phases. In the initial phase of landfill lifetime, these gases are released from porous structure of waste, in the later stages they migrate to landfill with air intrusion due to negative balance between the values of internal and external pressure. Nitrogen can be also generated in denitrification process.

Marginal components of landfill gas are hydrogen (H_2) and ammonia (NH_3). Hydrogen is a product of acid fermentation of organic matter. Its concentration depends on landfill life phase, and it is the highest in acetogenic phase. Ammonia is produced by ammonification, nitrogen-rich organic materials (e.g. protein—containing waste). This process is complex. In the initial phase the enzymatic hydrolysis of proteins leads to amino acids production. These compounds undergo deamination (amino group NH_2 is disengaged, resulting in formation of ammonia) or fermentation resulting in production of CO_2, ammonia nitrogen and volatile fatty acids (Berge et al. 2005). Ammonia concentration in LFG is usually not higher than a few ppm because the slightly acidic reaction of the waste favours its dissolution in the water, and formation of NH_4^+ ions, which pass into the leachate.

Due to the high moisture content of the waste and the temperature inside the bed reaching up to 50°C, the landfill gas is characterized by high water vapor saturation. Relative humidity of landfill gas is usually near the saturation point.

Beside of these small components landfill gas also contain more than 550 compounds, belonging to different chemical groups (Parker et al. 2002). They include both inorganic and organic compounds. Scott et al. (1988) divided trace constituents of biogas to 12 groups (cited by Parker et al. (2002): hydrogen sulphide (H_2S), alkanes, alkenes, cycling organic compounds, halogenated compounds, alcohols, esters, carboxylic acids, amines, ethers, organo-sulphur compounds, and other oxygenated compounds, such as ketones or furan derivatives. These compounds, apart from hydrogen sulphide are classified as volatile organic compounds

(VOCs) being the chemicals with initial boiling point less than or equal to 250°C measured at a standard atmospheric pressure of 1013 hPa (Directive 2004/42/EC).

All volatile organic compounds, except for methane, are called non-methane volatile organic compounds (NMVOCs). Their total content in the landfill gas ranges from 0.01 to 1.0 vol.% (Lauber et al. 2006) and depends mainly on the morphology of wastes, the age of the landfill and the season. This group includes, among others (starting from the highest concentrations): aromatic hydrocarbons, halogens, ketones, aliphatic hydrocarbons, esters, alcohols, cyclic aliphatics and aromatic hydrocarbons, sulphur compounds, siloxanes, aldehydes, organic acids (Jaffrin et al. 2003).

Among aromatic compounds, the ones that appear in the highest concentrations are benzene and its derivatives, toluene and xylenes. The presence of toluene was found in biogas immediately after the deposit of waste, due to high volatility of this compound. The aromatics content in biogas increases with the intensification of anaerobic conditions (Parker et al. 2002). According to Allen et al. (1997), the concentrations of aromatic compounds in landfill gas range from 94 to 1906 mg m⁻³.

Large group among NMVOCs is composed of halogenated volatile organic compounds (chlorinated and/or fluorinated). Most of these are one or two carbon organic compounds having a low molecular weight, such as carbon tetrachloride, chlorobenzene, chloroform, trifluoromethane, vinyl chloride, bromodichloromethane and chlorodifluoroethane. They are often present in landfill gas; however, they are only rarely formed in biogas during the anaerobic digestion of sewage sludge. Most halogenated volatile compounds are directly released from the waste due to direct volatilization, and their quantity depends on the internal conditions such as the pressure and the temperature in the waste body. Their concentration in LFG from UK landfills varied from 327–1239 mg m⁻³ (Allen et al. 1997). They originate from consumer products of the chemical industry used as refrigerants, propellants, and in insulating foams. Their stability in the atmosphere is very variable, from <1 year (e.g. methyl bromide, CH_3Br) to 1700 years (CFC-115, $CClF_2CF_3$). During the combustion they are oxidized to corrosive products or in favourable conditions they can initiate the formation of dioxins and furans (Persson et al. 2006).

Aliphatic hydrocarbons: alkanes (C_nH_{2n+2}), apart from methane and alkenes (C_nH_{2n}), are present in biogas in the greatest concentration in the early stages of waste decomposition, mainly taking place in aerobic conditions. They are poorly soluble in water and have low molecular weights, and therefore, they easily migrate into biogas. Alkanes with higher molar masses, such as nonane and decan are also present in the gas formed in the anaerobic phases. Alkenes concentrations are lower than alkanes (Parker et al. 2002). The concentration of alkanes in the gas emitted from seven municipal landfills in the United Kingdom varied in the range of 302–1543 mg m⁻³ (Allen et al. 1997).

Cyclic organic compounds: alicyclic cycloalkanes and cycloalkenes and aromatic hydrocarbons constitute a group of gases which are very toxic. Cycloalkenes (e.g. limonene, α—pinene) predominate the concentration of cycloalkanes (e.g. cyclobutane, cyclohexane). The concentration of the cycloalkanes in the gas formed in landfills in the United Kingdom ranged from 80 to 487 mg m⁻³ (Allen et al. 1997).

The volatile compounds include also volatile organosilicon compounds, among them siloxanes. Siloxanes are a group of Linear (L) and Cycling (D) organic compounds that contain silicon, oxygen and methyl groups. Their concentrations are generally higher in digester gas than in landfill gas. For instance, the silicon concentration in landfill gas in Germany usually varies between 1 and 8 mg m⁻³ (Arnold, 2009 cited by Häusler & Schreierin, 2005), while the average siloxane concentration in biogas produced in the German wastewater treatment plant was 14.9 mg m⁻³ (Arnold, 2009 cited by Beese, 2007). Contrary to digester gas, which contains mainly D4 (octamethylcyclotetrasiloxane) and D5 (decamethylcyclopentasiloxane), landfill gas contains significant quantities of D3 (hexamethylcyclotrisiloxane) and D6 (dodecamethylcyclohexasiloxane), as well as linear siloxanes, such as L2–L5 (hexamethyldisiloxane, octamethyltrisiloxane, decamethyltetrasiloxane dodecamethylpentasiloxane

(Wheless & Pierce, 2004). The concentration of siloxanes in gas from old and closed landfills is generally smaller than in new ones, where silicon-containing waste is continuously deposited. Siloxanes originate from some consumer products, such as shampoo, detergents and cosmetics. Presence of these compounds in LFG negatively affects the performance of energy generation devices, thus they should be removed before combustion. During combustion they are converted to silicon dioxide (SiO_2), which can deposit on valves, cylinder walls and liners are the cause of extensive damage by erosion or blockage.

Sulfur compounds, both inorganic and organic are mainly responsible for the odor nature of landfill gas. They are characterized by very low values of odor detection threshold in comparison with other compounds (Table 2.2). These include: sulphides (e.g. hydrogen sulphide, carbonyl sulphide COS), disulphides, thiols (mercaptans), sulphate and sulphite, and oxidized sulfur compounds. The main sulfur compound in biogas is hydrogen sulphide (H_2S). The reason for its high concentration in biogas is a high content of calcium sulfate in waste (Fairweather & Barlaz, 1998). Bacteria, such as *Desulfovibrio* and *Desulfotomaculum* use sulfates as the electron acceptor released during the oxidation of organic matter, converting them to hydrogen sulfide. Concentration of H_2S in LFG is in range of 100–1000 ppm (Table 2.1), but can rise even to 1.2 vol.% (Lee et al. 2006) in landfills, where crushed construction waste, containing gypsum fiber panels have been deposited or used as cover material, or at landfills with large deposits of plasterboards or sulphate containing sludge, such as sewage sludge or flue gas desulphurization sludge. A portion of H_2S migrates into biogas, whereas the majority of it dissolves in water contained in waste and as dissolved H_2S or HS^- ion, it enters leachate (Erses & Onay, 2002).

The highest concentration of hydrogen sulphide in biogas is noted in the early stages of waste decomposition. The decrease in the concentration of H_2S is most likely caused by the precipitation of the sulphides in the reaction with heavy metals (such as Cu and Fe) or their oxides, which are present in the deposited material. Sulphides as water insoluble compounds remain in the mass of waste (Parker et al. 2002). The organic sulfur compounds in the greatest concentrations in landfill gas are: dimethyl sulphide (DMS), carbon disulphide, methyl mercaptan, dimethyl disulphide (DMDS) at the concentrations of: 0.007–180 mg m^{-3}; 0.09–61.6 mg m^{-3} 0.084–17.94 mg m^{-3}; 0.0124–0.942 mg m^{-3}, respectively (Kim et al. 2005, Shin et al. 2002).

In the early stages of degradation the following compounds appear in biogas: alcohols, which are intermediate products of decomposition of the complex organic matter to carboxylic acids; esters resulting directly from the decomposition of waste or from the reaction between the carboxylic acids and alcohols, they are highly odorous compounds of a sweet smell; carboxylic acids, ethers, amines (such as dimethylamine with fishy odor) and ketones and furans. The concentration of alcohols and ketones in LFG arising in landfills in the United Kingdom ranged from 2 to 2069 mg m^{-3} (Allen et al. 1997).

The range of concentrations of selected volatile organic compounds found in landfill gas are presented in Table 2.3.

Table 2.2   Odour threshold value of selected trace gases (Guidance of Landfill Gas Flaring, 2002).

Compound	Odour threshold value [µg m^{-3}]	Compound	Odour threshold value [µg m^{-3}]
Hydrogen sulphide	0.1	Carbon disulphide	700
Dimethyl disulphide	4	Toluene	700
Dimethyl sulphide	3.7	Trichloroethylene	3000
Limonene	20	Methanol	6000
Phenols	20	Benzene	9000
Naphthalene	50	Propanol	10000
Styrene	70	Chloromethane	21000
Xylenes	540	Ethanol	100000

Table 2.3 Concentrations [mg m^{-3}] of selected NMVOCs in landfill gas.

NMOCs group	Chemicals	EPA 1995 (converted from ppm)	Shin et al. 2002 (converted from ppm)	Jaffrin et al. 2003	Lauber et al. 2006 (converted from ppm)	Kim et al. 2006 (converted from ppb)	Rasi et al. 2007	Pawłowska et al. 2008
Aromatic hydrocarbons and chlorinated aromatic hydrocarbons	Benzene	1.97	23–37.8		7.22	0.001–9.15	0.6–2.3	21–40
	Toluene	59.11	88.4–182	65–87	142	0.015–131.1	17–51	22–94
	Ethylbenzene	22	21–58.4		32.8	0.012–38.02		<117
	m-Xylenes	27.9	21.5–67.3	37.89	12.2	0.001–33.06		<32
	p-Xylene							<DL
	o-Xylene	10.1				0.001–9.442		<DL
	Styrene	9.26				0.17–21.99		
	Chlorobenzene	5.7						<380
Aliphatic hydrocarbons and chlorinated aliphatic hydrocarbons	Ethane	293.4						
	Propane	25.8						
	Dichloromethane				93.5			<58
	Chloroform (trichloromethane)		413–966.7					<72
	Carbon tetrachloride		281–841.9					
	Chloroethylene (vinyl chloride)				11			
	Trichloroethylene		5.21–15		12.1			
	Tetrachloroethylene		19–50.4		36.5			
Terpenes	α-Pinene	47.1						
	Limonene	212.2						
Ketones	Total			218.82				
	Acetone (propanone)	15.5		7.84	17.9			
Organic silicon compounds							0.7–4	<192
								<130

2.3 QUANTITATIVE ESTIMATION OF LANDFILL GAS PRODUCTION

Similarly to the composition, the amount of landfill gas varies with the age of the landfill. In general, the most intense release of gas begins after approximately two years from the landfill closing and lasts for about 20–30 years. The amount of gas generated at the landfill depends mainly on the amount of accumulated waste and its morphological composition, as well as the phase of biochemical transformation of waste.

The composition of waste directed to landfill depends on many factors, i.e. the original composition of produced waste, the way in which waste is treated, including the type of selective collection of source-separated waste or residual municipal solid waste collection (RMSW) and approved method of RMSW treatment. The initial morphological composition of municipal solid waste is variable, both in terms of spatial and temporal dimension. It mainly depends on the economic status of the society, the size and nature of human settlements (towns, villages), the type of housing infrastructure (one and multifamily housing; dense or dispersed), as well as technical and sanitary equipment of buildings. Table 2.4 shows an exemplary data concerning the morphological composition of municipal waste for selected countries in Europe and the world. The presented data reveal that the percentage of organic waste (except for cellulosic waste) in the total mass of waste ranges from 23.3% (the United States) to 64.8% (household waste—Hangzou, China). The share of paper and cardboard, which are also biodegradable, although much slower, ranged from 6.71% (Hangzou, China) to 37% (Finland). However, these data are just approximate. Application of various research methodologies and a relatively large time interval from which the quoted data hinders the comparison of the results of examination of waste composition. The methodological approach to the study of the morphological composition varies all over the world (Dahlén & Lagerkvist, 2008). For instance, in the United States the methodology of ASTM is mainly used, whereas in Austria, Germany, Italy, Spain, United Kingdom, Romania and Poland, researchers use the methodology SWA-Tool, which so far is not mandatory in all EU countries. Moreover, many member states use the methodology MODECOM, Nordtest or ARGUS.

Changes in the morphological composition of waste that have taken place over time are showed by the studies conducted in the years 1992–2004 in the Polish city Wroclaw, with the population of about 630 thousand residents (Mackow et al. 2005). In this period, the decline in the share of fines, and the increase in glass and plastic content were observed. Morphological composition of waste also depends on the season. Boldrin & Christensen (2010) presented that the amount of garden waste per capita varied from 2.5 kg month[-1] in winter, up to 19.4 kg month[-1] in summer.

Nowadays, the most important factor affecting the morphology of the landfilled waste in countries that implement "waste minimization strategy" is the way of their pretreatment. In many countries, as a result of adopted environmental policy, which aims at reducing the negative impact of waste management sector on the environment, a system of selective collection of waste is being promoted. Separation from the waste stream the fractions that can be recycled affects the increase of the percentage of the remaining organic matter in waste to the amount from 63 to 90% (De Baere, 2000, Cecchi et al. 2002). Wastes (called as RMSW), rich in organic matter, are also produced by a mechanical process, used in the mechanical-biological treatment (MBT) plant. Processes of wastes shredding and sieving lead to the separation of the fine fraction, usually with a diameter below 70 mm, so called mechanically sorted organic fraction (MSOR). High volatile solids content, usually exceeding 50% and moisture content in the range 40–55% by weight (Woelders et al. 1993, De Gioannis et al. 2009) determine the necessity of subjecting the wastes to stabilization processes before placing them into the landfill. Without the stabilization of these wastes, they have a high susceptibility to microbial degradation, which is associated with the high biogas potential in anaerobic conditions. The model tests conducted by Pan & Voulvoulis (2007) reveal that by using the GasSim model for estimation of biogas production, only 50–60% reduction of

Table 2.4 Composition of municipal solid waste from different countries of the world.

Waste type	Country, city, year	Waste component												References
		Organic waste	Wood	Paper and cardboard	Plastics	Glass	Textiles/rubber/leather	Metals	Hazardous waste	Multi-components	Inert waste	Fine fraction <10 mm	Others	
Municipal solid waste (MODECOM)	Poland, Poznań, 2001	23.7	–	14.1	10.8	9.2	2.9/-/-	2.1		1.5		20.1	15.6[1]	den Boer et al. (2009)
Municipal solid waste (SWA Tool)	Poland, Kraków, 2003	40.5	0.9	10.2	12.1	10.1	2.7	1.8	0.2	3.1	5.6	8.5	4.1	Jędrczak & Szpadt (2006)
Municipal solid waste	Poland, Grudziądz, 2000	29.4	–	17.3	3.9	4.8	10.6/-/-	1.2				21.7		Jędrczak & Pilicydis (2000)
Municipal solid waste (ASTM)	USA, 2001	23.3	7.4	28.0	14.9	6.3	8.5	7.4				–	4.2	US EPA (2003)
Municipal solid waste	Singapur, 2000	41.53	8.9	20.6	5.8	1.1	0.9/-/0.2	3.2			4.3	–	17.8[2] 4.51	Bai & Sutanto (2002)
Household waste	Hangzhou, China	64.48	0.05	6.71	10.12	2.02	1.22	0.31	0.05			–	15.04	Zhuang et al. (2008)
Municipal solid waste (AFNOR, XP X30–408)	Mende Lozére, France	9.1	–	23.3	14.8	4.2	3.2/-/-	5.4	1.1	3.6	2.9	20.5	12.1[3]	Bayard et al. (2010)
Municipal solid waste	Granada, Spain	30.5	1.5	24.0	21.0	12.0	1.0	5.0						5.0	Zamorano et al. (2007)
Municipal solid waste	Verona, Italy, 1998	27	–	30	15	9	–	4						15	Bolzonella et al. (2003)
Municipal solid waste	Thiruvananthapuram, India, 2005	50.5	–	10.5	7.6	2.4	2.6	2.2		8.4				14.6[4]	Narrayana (2009)
Municipal solid waste	Weles, Macedonia	32.69	–	24.48	7.0	7.19	–	6.09						23.36	Hristovski et al. (2007)
Municipal solid waste	Finland	26.0	–	37.0	9.0	8[5]	1.0	–						19.0	Pipatti & Savolainen (1996)
Municipal solid waste	Poland														Landrat (2007)
a) from urban area		40	–	22.6	7.3	–	–	–				–	–		
b) from rural area		23	–	3.5	1.0	–	–	–				–	–		

[1] sanitary products (2.7%), combustible mixed waste (3.5%), non-combustible mixed waste (4.5%), special waste (1%), sorting loss (4%), used slag (4.3%), construction debris (4.5% earth spoils (2.7%), sludge (1.8%), others (4.5%), [2] sanitary waste (8.4%), non-classified combustible (3.7%); [3] ash (3.8%), sand (10.8%); [4] glass and metals.

organic matter content in MSOR obtained under biological treatment allows to gain the properties of MSOR only on the level similar to typical MSW. Therefore, the intensive stabilization is a necessary stage preceding the MSOR deposition in the landfill.

Making the methods of mechanical-biological waste treatment more popular will significantly change the composition and properties of landfilled waste. This will reduce the production of landfill gas and leachates, and will influence their composition. According to Stegmann (2005) biogas generation from waste after MBT will be in range from 0 to 20 m³ Mg⁻¹ of total solids, which is equivalent to more than 90% reduction in the production of biogas in relation to raw MSW. The reduction level of biogas generation depends on the solution of pretreatment of waste. According to the research carried out by De Gioannis et al. (2009), aerobic stabilization of mechanically pretreated RMSW, lasting 8 and 15 weeks, decreased the landfill gas generation potential by 83% and 91%, shortened the non-methanogenic phase by 67% and 82%, and decrease the amount of gasified organic carbon by 81% and 93%, respectively to aerobic stabilization time, compared to the raw waste. Scaglia et al. (2010) also observed significant reduction of biogas potential, which equaled 56% and 79% after 4 and 12 weeks of aerobic treatment of mechanically screening MSW, respectively. Examinations of Mahar et al. (2009) showed that leachate recirculation during the aerobic stabilization of residual waste fraction enhance the process. They stated that the CH_4 emissions from the waste conditioned by natural convection of air, natural convection combined with leachate recirculation, and forced aeration combined with leachate recirculation, were reduced about 26%, 45% and 55%, respectively, as compared to anaerobic condition. Aerobic stabilization can also be successfully applied for unsorted waste, only pretreated by shredding. Forced-aeration process carried out by 25 weeks, reduced the biogas production potential of these waste by 90% and the mass of waste going to landfill by 28% (Lornage et al. 2007). Therefore, waste pretreatment is a very effective way to reduce gas emissions from landfills.

Time dependent quantitative estimation of LFG production is the primary criterion that should determine the choice of method of gas utilization. This assessment is mostly performed with the use of mathematical models, based on data concerning the amount and the composition of accumulated waste, the time of their collection, and assigned or assumed parameters of their distribution (e.g. degradation rate constant, conversion factor). Models allow for the estimation of annual biogas/methane production [m³ Mg⁻¹ year⁻¹]. After addition of these data the cumulative biogas/methane production [m³], that is the total volume of biogas/methane produced at the particular landfill during its full biochemical activity, is estimated. The accuracy of estimates of production or emission of landfill gas depends on the type of the model adopted. It is testified by the results of model tests conducted by Scharff & Jacobs (2006). They found that the emission of landfill gas, calculated with the use of six different models for three landfills in the Netherlands were extremely different. The highest estimates were five to seven times higher than the lowest estimates. It results mainly from the construction of the models that, although they assume that the decomposition of wastes follows by the first order kinetics, they still take into account the various input parameters. Most of the models does not allow for variations in composition of waste in terms of morphology and susceptibility to degradation, provided that the entire mass of organic matter in waste is degraded at the same rate. More complex models distinguish between different waste category (multi-phase model Afvalzorg, first order model TNO), and even fractions that are characterized by different biodegradation constant rate (multi-phase model Afvalzorg).

The basic condition to obtain the results of the modeling, which would be as suited as possible to the actual production of gas is the right choice of assumptions, such as methane generation potential or degradation rate constant. These parameters are strongly dependent on the chemical composition, properties of waste and the conditions of the process. Therefore, they should be selected individually for each object. Methane/biogas generation/production potential is defined as the amount of methane/biogas that can be achieved with the mass of waste and is expressed in, for instance, m³ Mg⁻¹. In case of a mixture of waste deposited in the landfill, their potential is a resultant value of the potentials of particular waste categories, which are significantly different from each other (Table 2.5). This potential can be determined theoretically or on the basis of experiments carried out in a laboratory scale (in batch or continuous flow systems), or on the basis of data obtained in an industrial scale.

The theoretical biogas production per unit weight of the organic compounds may be calculated on the basis of the elemental composition of digested waste. In order to calculate the amount of the most important components of biogas produced during the digestion of 1 mole of organic matter with a specific elemental composition, contained in waste, modified Buswell equation [Equation 2.1] (Nyns, 1986) can be used.

$$C_a H_b O_c N_d + \left(\frac{4a - b - 2c + 3d}{4} \right) H_2O \rightarrow \left(\frac{4a + b - 2c - 3d}{8} \right) CH_4$$
$$+ \left(\frac{4a - b + 2c + 3d}{8} \right) CO_2 + dNH_3. \tag{2.1}$$

Theoretical biogas production potential of waste ranged from 150 to 265 m^3 Mg^{-1} (Bhide, 1990; Zamorano et al. 2007). Such wide range of results is caused by the differences in proportions of C:H:O:N in the molecular formula of waste, which depends on morphological composition of waste. The biogas production potential of municipal solid waste from Granada (Spain), described by the formula of organic matter $C_{44}H_{70}O_{29}N$ was calculated for 160 m^3 Mg^{-1} raw waste (Zamorano et al. 2007). Theoretical biogas production potential of food wastes coming out from hazelnut, rice, and wine industrial processing, described by the molecular formulas: $C_{49}H_{53}O_{31}N$, $C_{100}H_{125}O_{43}N$ and $C_{33}H_{58}O_{28}N$ were 960 m^3 Mg^{-1}VS; 790 m^3 Mg^{-1} VS, and 900 m^3 Mg^{-1} VS, respectively (Roati et al. 2012).

Molecular formula of organic matter contained in waste depends on the content of lipids, carbohydrates, proteins, and others compounds. According to stoichiometry, the highest biogas production is obtained from fats. Theoretically, in normal conditions (0°C and 1 atm.), there is about 1444 m^3 of biogas, containing 70.2% CH$_4$ created from 1 Mg of lipids (trioleic acid); whereas from 1 Mg of cellulose it can be obtained 830 m^3 of biogas, containing 50% CH$_4$; and from 1 Mg of protein—793 m^3 of biogas, containing 63.6% CH$_4$ (Jørgensen, 2009). The content of individual chemical compounds in the composition of organic waste depends on the morphological composition of waste. A large share of yard waste and food waste is associated with a high content of cellulose and hemicelluloses. Holocellulose (cellulose and hemicelluloses) is the important degradable component of MSW (Baldwin et al. 1998), which is responsible for 91% of methane potential of the waste (Barlaz et al. 1989).

In practice, not all of the organic matter in waste is subject to microbial degradation. It makes the production of landfill gas is much lower than the theoretical value, calculated from equation 2.1. At Polish landfills, there is only from 60 up to 120 m^3 of biogas obtained from 1 Mg of municipal waste (Ciupryk & Gaj, 2004). Not only the concentration of biodegradable compounds in waste, but also their biodegradability determine the quantity of gas produced. It is generally accepted that food waste, paper and cardboard are completely biodegradable; whereas textiles and mixed non-combustible materials are biodegradable only in 50%, and when it comes to the combustible mixed waste—in 60% (Burnley, 2001).

Some wastes contain large amounts of lignin, which is considered to be only a little bit susceptible or completely resistant to biodegradation under anaerobic conditions (Eleazer et al. 1997, Sanders et al. 2002, Fox & Noike, 2004). Lignin constitute from a few to several percent of dry weight of MSW (Jones & Grainger, 1983, Barlaz et al. 1990, Rao et al. 2000, Barlaz, 2006). According to Tchobanoglous et al. (1993) among the household waste, newsprint and cardboard contain the biggest amount of lignin (21.9 and 12.9% of VS, respectively), whereas the smallest amount is in food waste (0.4% of VS). Taking into account the data obtained in the study on anaerobic digestion of lignocellulosic materials, Chandler et al. (1980) proposed that the share of biodegradable fraction in mass of organic compounds should be calculated on the basis of following linear relationship (Equation 2.2):

$$BF = 0.83 - (0.028 \cdot X_l) \tag{2.2}$$

where BF = biodegradable fraction expressed on the volatile solids basis ($0 < BF < 1$); 0.83 and 0.028 = empirical constants; X_l = initial lignin content as a percent of volatile solids [%].

Potential of biogas/methane generation of waste could be also determined experimentally in batch tests, semi-continuous-flow or continuous-flow reactors. Batch tests are based on measurement of the amount of biogas (and the concentration of CH_4 in biogas) isolated during the fermentation of the portion of material placed in a closed vessel. Parameters determined in the batch test are called biogas potential or Biochemical Methane Potential (BMP). A detailed review of conditions in which these tests are conducted is presented in the work made by Angelidaki et al. (2009).

In continuous and semi continuous-flow systems waste go into reactor continuously or at certain intervals. The introduction of a new portion of feedstock is accompanied by a simultaneous discharge of an equivalent quantity of the batch of reactor. Part of the material leaves the reactor without being completely decomposed. Parameters determined in these systems are called biogas/methane yield. Some results of laboratory estimation of biogas production from waste are presented in Table 2.5.

The potential of biogas/methane was studied both for mixed waste as well as their particular biodegradable components, such as paper, cardboard or kitchen waste. Biogas potential of mechanically pretreated municipal solid waste, examined in the batch test ranged from 330–400 m³ Mg⁻¹ of VS (Braun et al. 2003, Scaglia et al. 2010). A fraction of organic waste from households (food waste, vegetable and fruit waste) is very susceptible to biodegradation and is characterized by the greatest biogas potential reaching 808–813 m³ Mg⁻¹of VS (Schievano et al. 2009). Significantly lower potential (from 62 to 288 m³ Mg⁻¹ of TS, calculated assuming 50% of CH_4 in biogas) is observed in case of green waste (leaves, branches, grass) (Barlaz et al. 1997).

Biogas yield calculated for the data obtained in technical scale for kitchen wastes ranged from 460 to 880 m³ Mg⁻¹ of VS, for garden waste from 152 to 426 m³ Mg⁻¹ of VS, and for paper and cardboards from 256 to 490 m³ Mg⁻¹ of VS, depending on the technology used (Chavez-Vazquez & Bagley, 2002).

Apart from the parameter assessing the biogas generation potential, other important parameter required for the modelling of LFG production is the first-order degradation rate constant (k), describing the degradation rate. In popular models of estimating the production of biogas/methane in landfill, the proposed value of the constant k ranges from 0.013 y⁻¹ (the lowest value in the GasSim multi-phase model) to 0.231 y⁻¹ (multi-phase model Afvalzorg). The k-values obtained for three landfill in the Netherlands were in range from 0.03 y⁻¹ (for slowly degradable waste), through 0.099 to 0.116 y⁻¹ for moderately degradable, and 0.187 to 0.231 y⁻¹ for rapidly degradable waste (Scharff & Jacobs, 2006). Value of degradation rate constant depends mainly on chemical composition of waste.

2.4   LANDFILL GAS UTILIZATION

Landfill gas is usually slightly lighter than air which promotes its release into the atmosphere. Density of LFG lowers with the increase of CH_4 concentration, which is colourless and combustible gas that burns with blue flame (Itodo et al. 2007). Due to the presence of harmful substances, flammability and odour character, the aim is to reduce LFG emissions to the environment, subjecting it to utilization. The choice of the method of landfill gas utilization is influenced by its quality and quantity (Figure 2.1). When the concentration of CH_4 in LFG is in the range of 35–40% and the output exceeds 30 m³ h⁻¹, it is technically

Table 2.5 Biogas potential/biochemical methane potential of municipal solid waste of their components.

Waste type	Experiment conditions	Time of the experiment or hydraulic retention time (HRT) (days) and organic loading rate (OLR)	Biogas potential (B)/ biochemical methane potential (CH_4) or methane yield	Reference
Source separated municipal biowaste Biowaste (31%) + Sewage sludge (69%)	Laboratory scale, batch test, mesophilic condition (35°C)	27 30	0.4 m³B kg VS⁻¹ 0.54 m³B kg VS$_{add}$⁻¹	Braun et al. (2003)
Food wastes from fruit and vegetable markets, households, hotels and juice centres,	Laboratory scale, batch test ($V = 3.25$ dm³), mesophilic conditions (26 ± 4°C)	100	Max. 0.661 m³B kg VS⁻¹	Rao et al. (2000)
Municipal solid waste after MBT (fraction <60 mm),	Laboratory scale, batch test (reactor volume 0.5 dm³)	15	0.204 ± 0.082 m³ B kg TS⁻¹ $= 0.33 \pm 0.13$ m³ B kg VS⁻¹	Scaglia et al. (2010)
Source separated household waste	Laboratory scale, batch test (reactor volume 0.1 dm³), mesophilic conditions	60	0.777–0.782 m³ B kg TS⁻¹ $= 0.808$–0.813 m³ B kg VS⁻¹	Schievano et al. (2009)
Mechanically pre-treated MSW Food waste	Laboratory scale, batch test, $V = 1$ dm³, thermophilic conditions,	20	0.482–0.522 m³ B kg VS⁻¹ 613 m³ B kg VS⁻¹	Zhu et al. (2009)
Grass Leaves Branches Food waste Office paper	Laboratory scale, batch test (reactor volume 2.0 dm³), mesophilic conditions	80	0.144 m³ CH_4 kg TS⁻¹ 0.031 m³ CH_4 kg TS⁻¹ 0.062 m³ CH_4 kg TS⁻¹ 0.300 m³ CH_4 kg TS⁻¹ 0.217 m³ CH_4 kg TS⁻¹	Barlaz et al. (1997)
Putrescible fraction of food market waste	Laboratory scale semi-continuous reactor, $V = 3$ dm³ mesophilic conditions (35°C)	HRT 8–20 OLR <3 kg VS m⁻³ d⁻¹	0.478 m³ CH_4 kg VS$_{add}$⁻¹	Mata-Alvarez et al. (1992)
Source separated municipal biowaste	Laboratory scale, semi-continuous reactor, $V = 4,5$ dm³ wet fermentation, thermophilic conditions (55°C), leachate recirculation	HRT 14–18	0.63–0.71 m³ B kg VS⁻¹	Hartmann & Ahrig, (2005)
Paper (47.3%), cardboard (11%), plastics (11%), household biowaste (5.9%) Paper (91.5%), cardboard and household biowaste	Pilot scale, three phases fermentation with leachate recirculation, thermophilic conditions (55°C)	HRT 21, 42 OLR 3, kg VS m⁻³ d⁻¹ HRT 21 OLR 6.4 kg VS m⁻³ d⁻¹	0.33–0.4 m³ B kg VS⁻¹ (42 days) $= 0.18$–0.22 m³ CH_4 kg VS⁻¹ 0.22–0.35 m³ B kg VS⁻¹ (21 days) $= 0.13$–0.19 m³ CH_4 kg VS⁻¹ 0.28–0.35 m³ B kg VS⁻¹ (21 days) $= 0.17$–0.22 m³ CH_4 kg VS⁻¹ CH_4 concentration: 55–60 vol.%	Chynoweth et al. (1992)
Mechanically sorted organic fraction of MSW + putrescent fraction from the source sorted MSW	Pilot scale, thermophilic conditions, semi-dry	HRT 13.5 OLR 9.2 kg VS m⁻³ d⁻¹	0.23 m³ B kg VS⁻¹ 0.16 m³ CH_4 kg VS⁻¹ CH_4 concentration: 68.7 vol.%	Bolzonella et al. (2003)
Food waste	Batch scale $V = 3.6$ dm³ Semi-continuous feed tests, $V = 30$ dm³	1. 20 2. HRT 30 OLR 1.5 and 3.0 kg VS m⁻³ d⁻¹	0.15–0.56 m³ CH_4 kg VS⁻¹ (batch) 0.313 and 0.321 m³ CH_4 kg VS⁻¹ (semi-continuous test)	Qamaruz-Zaman & Miike (2010)

feasible and economically justified to use gas as a source of heat or electricity (Haubrichs & Widmann, 2006). From the point of view of sustainable development, it is the most benefi-cial option of LFG utilization because at the same time the "green" energy from alternative sources is obtained and the emission of methane and other pollutants into the atmosphere is prevented.

Considering the use of LFG for the energy production, the following physical parameters should be determined: lower and higher heating values, Woobe index, methane number, auto-ignition temperature and flammability limits. The lower heating value (LHV), corresponding to the energy that is released when one normal cubic meter (Nm^3) of biogas is combusted and the water vapour is not condensed (the heat of condensation is not included), varies from 12 to 22 MJ m_n^{-3}. Decrease in CH_4 concentrations of 5% lowers the LHV of about 1.6 MJ m^{-3} (Czurejno, 2006). Higher heating value (HHV) of biogas including the heat of condensation ranges from 18–30 MJ m_n^{-3}. The heating values fall with a decrease of CH_4 concentration in landfill gas.

Biogases from different sources have very high methane number (MN), more than 130 (Persson et al. 2006, Malenshek & Olsen, 2009). This parameter describes the gas fuel resist-ance to knocking during a combustion. An assessment of its value bases on a comparison of knock rating of gaseous fuel and reference fuels that are the mixtures of hydrogen and methane. Methane number for CH_4 is 100 and for H_2 is zero. Methane number depends not only on CH_4 and H_2 content, but also on volume fraction of inert gases such as CO_2 and N_2. Methane number lowers with the decrease of CH_4 content and it grows with the increase of volume fraction of CO_2 and N_2. The higher methane number the greater knocking resistance of fuel. It is assumed that gases considered as good and very good applicable for combus-tion in gas engine have a MN > 85 (Major, 1993). Wobbe index, being the indicator of the interchangeability of gases used as fuels, is about 3 times lower for LFG than for natural gas. That means that it cannot be introduce into the gas grid without upgrading. Wobbe index is defined as the ratio of higher heating value (HHV) and square root of specific gravity (called relative density) of the gas. Wobbe index limit values applicable in the case of injection of biogas into the gas grid vary across EU countries. They are in a wide range of 43.74–54.70 MJ m_n^{-3} (Williams, 2009). In case of low calorific gases the determinant factor of the useful-ness of fuel to combustion in definite device, is the flame speed. It must be within the range specified by producer. The flame speed of biogas is low, just 25 cm s^{-1} as against 34 cm s^{-1} for natural gas and 275 cm s^{-1} for hydrogen.

High auto-ignition temperature of LFG, reaching about 650° resists knocking, which is desirable in spark-ignition engines (Porpatham et al. 2008). Biogas has narrower flammabil-ity limits than natural gas (7.5–14 vol.%). Flammability limits refer to the range of composi-tions, for fixed temperature and pressure, within which an explosive reaction is possible when an external ignition source is introduced.

The comparison of physical properties of biogases with various origins and natural gas are depicted in Table 2.6. Biogases have lower calorific value, flame speed and flammability limits compared with natural gas, mainly because of CO_2 presence (Porpatham et al. 2008).

Despite the fact that landfill biogas is produced over decades, its production with parame-ters allowing for the use of energy in the sanitary landfill in optimal condition lasts for 20–30 years (Huber-Humer et al. 2008). Thus, the period when landfill generates a low calorific gas lasts much longer than the time of using biogas as energy source (Haubrichs & Widmann, 2006). When the concentration of CH_4 decreases to a value in the range 20–25%, and the production drop to 10–15 m^3 h^{-1}, gas may be combusted in high temperature flare. Below the mentioned values, the utilization of biogas may be carried out by its combustion with the auxiliary gas, by catalytic combustion, preferably by using Pd-based catalysts (Smith et al. 2006) or non-catalytic oxidation in high temperatures (Stachowitz, 2001). However, these methods are expensive. Biotechnological methods, which are based on the microbiological oxidation of contaminants (methane and trace constituents of biogas) during the gas flow through the filter bed, are much cheaper and easier to use. It is assumed that they provide an

Table 2.6 Physical parameters of landfill gas, biogas from anaerobic digesters and natural gas.

Parameter	Unit	Landfill biogas	Anaerobic digesters	Natural gas	References
Lower heating value	MJ m_n^{-3}	12–22			Kalina & Skorek (2002)
	MJ m_n^{-3}	16	23	40	Persson et al. (2006)
Density	kg m_n^{-3}	1.3	1.2	0.84	Persson et al. (2006)
Wobbe index, upper	MJ m_n^{-3}	18	27	55	Persson et al. (2006)
Methane number		>130	>135	70	Persson et al. (2006)
		139.6	139.1		Malenshek & Olsen (2009)
Auto-ignition temperature	°C		650*	540	Porpatham et al. (2008)
Flammability limits	vol.% in air		7.5–14*	5–15	Porpatham et al. (2008)
Flame speed	cm s^{-1}		25*	34	Porpatham et al. (2008)

*unknown biogas origin.

effective treatment of the gas, wherein the methane content does not exceed 5%. However, using aeration the CH_4 concentration may be increased up to 15% (Stachowitz, 2001).

2.5 SUMMARY

The composition of biogas, determining its properties, and its quantity are closely related to the morphology of the deposited waste and the phase of transformations that occur in them. This phase is related to the age of the landfill. Time-dependent variability of the composition and the amount of biogas produced hinders the use of one method of reducing greenhouse gas emission relevant for all landfills, regardless of their age. Therefore, it is insisted on developing long-term strategies for dealing with gas, aiming at the reduction of the uncontrolled release in various stages of landfill lifespan. When it comes to the choice of the strategy, it should be guided by the technical conditions of the landfill (capacity, surface area, and location), waste morphology, landfill age, climatic conditions, but also economic and binding law regulations of particular country. The latter sometimes imposes the directions of the proceedings. The strategy adopted in the European Union is based on limiting the disposal of organic matter. Waste going to the landfill of non-hazardous waste shall not contain more than 5% of total organic carbon (TOC). Achieving this goal is only possible through the dissemination of selective waste collection and the increase of the efficiency of mechanical-biological pre-treatment of mixed waste before their disposal, which will help to reduce the production of landfill gas. This will prevent or significantly reduce the possibility of the use of LFG in energy production, but will not eliminate the risks stemming from uncontrolled biogas emissions into the environment. However, the prominent decrease of biogas will promote the applications of biological methods, such as biofiltration, in landfill gas mitigation.

In contrast to the strategy developed in the European Union, in the United States there is a concept of anaerobic bioreactor landfill which has been promoted over many years. Its aim is to reduce the impact on the environment through the intensification of biogas production and its usage in energy. In accordance with this solution, the landfill waste should be rich in easily biodegradable organic matter, preferably after the initial separation of the fractions not susceptible to degradation. The creation of optimal conditions for the development of microorganisms responsible for the anaerobic decomposition of organic matter (mainly to maintain high humidity) increases the rate of biogas production and reduces the time required to achieve stabilization of the waste mass. In comparison with the traditional dry sanitary landfill, this solution allows for better use of the energy potential of biogas and reduces the risk of uncontrolled LFG emissions.

Other strategies connected with diminishing the impact of waste disposal on the environment are based on changing the conditions inside the waste mass from anaerobic to aerobic. It changes the composition of the emitted gas and leachate, making them less dangerous for the environment. Aerobic conditions can be maintained throughout the whole landfill life (aerobic bioreactor landfill) or only in the phase ending with methanogenesis, in order to accelerate waste stabilization. Maintaining aerobic conditions which are artificially produced is very expensive, and therefore, these methods are not often used. A partial solution to the problem of LFG emissions is the use of semi-aerobic bioreactor landfill system. It based on air migration to the waste layer which is forced by pressure differences.

Fundamentals of biotechnology, advantages and disadvantages of these methods of landfill gas mitigation are explored in the following chapters of this book.

REFERENCES

Allen, M.R., Braithwaite A., Hills Ch.C. 1997. Trace organic compounds in landfill gas at seven U.K. Waste Disposal Sites. *Environ. Sci. Technol.*, 31(4): 1054–1061.

Andreottola, G. & Cannas, P. 1992. Chemical and biological characteristics of landfill leachate. In T.H. Christensen, R. Cossu, R. Stegman (eds.), *Landfilling of waste: leachate*. London and New York: Elsevier Applied Science.

Angelidaki, I., Alves, M., Bolzonella, D., Borzacconi, L., Campos, J.L., Guwy, A.J., Kalyuzhnyi, S., Jenicek, P. van Lier, J.B. 2009. Defining the biomethane potential (BMP) of solid organic wastes and energy crops: a proposed protocol for batch assays. *Water Sci. Technol.* 59(5): 927–934.

Arnold, M., *Reduction and monitoring of biogas trace compounds*. VTT Technical Research Centre of Finland, Research Notes 2496, 2009 (http://www.vtt.fi/inf/pdf/tiedotteet/2009/T2496.pdf)

Bai, R., Sutanto, M. 2002. The practice and challenges of solid waste management in Singapore. Waste Manage., 22: 557–567.

Baldwin, T.D., Stinson, J., Ham, R.K. 1998. Decomposition of specific materials buried within sanitary landfills. *J. Env. Eng.* 124(12): 1193–1202.

Barlaz, M.A. 2006. Forest products decomposition in municipal solid waste landfills. *Waste Manage.* 26(4): 321–333.

Barlaz, M.A., Eleazer, W.E., Odle, W.S. III, Qian, X., Wang, Y.-S. *Biodegradative analysis of municipal solid waste in laboratory-scale landfills*. Project Summary. USEPA 1997.

Barlaz, M.A., Ham, R.K., Schaefer, D.M. 1989. Mass-balance analysis of aerobically decomposed refuse. *J. Env. Eng.* 115(6): 1088–1102.

Barlaz M.A., Ham, R.K., Schaefer, D.M. 1990. Methane production from municipal refuse: A review of enhancement techniques and microbial dynamics. *Critical Reviews in Env. Control* 19(6): 557–584.

Bayard, R., de Araújo Morais, J., Ducom, G., Achour, F., Rouez, M., Gourdon, R. 2010. Assessment of the effectiveness of an industrial unit of mechanical-biological treatment of municipal solid waste. *J. Hazard. Mater.* 175: 23–32.

Beese, J. Betriebsoptimierung der motorischen Gasverwertung durch den Einsatz von Gasreinigungsanlagen; Siloxa Engineering AG. Presentation at Deponiegas 2007 FH Trier Saksa 10.–11.1.2007.

Berge, N.D., Reinhart, D.R., Townsend, T.G. 2005. The fate of nitrogen in bioreactor landfills. *Crit. Rev. Env. Sci. Tec.* 35(4): 365–399.

Bhid, A.D, Gaikwad, S.A., Alone, B.Z. In: *Proceeding of International Workshop on Methane Emissions from Waste Management, Coal Mining and Natural Gas Systems*. 1990, USEPA Workshop.

Boldrin, A., Christensen, T.H. 2010. Seasonal generation and composition of garden waste in Aarhus (Denmark). *Waste Manage.* 30(4): 551–557.

Bolzonella, D., Innocenti, L., Pavan, P., Traverso, P., Cecchi, F. 2003. Semi-dry thermophilic anaerobic digestion of the organic fraction of municipal solid waste: focusing on the start up phase. *Bioresource Technol.* 86(2): 123–129.

Bozkurt, S., Moreno, L., Neretnieks, I. 2000. Long-term processes in waste deposits, *Sci. Total Environ.* 250: 101.

Braun, R., Brachtl, E., Grasmug, M. 2003. Codigestion of proteinaceous industrial waste. *Appl. Biochem. Biotech.* 109: 139–153.

Burnley, S. 2001. The impact of the European landfill directive on waste management in the United Kingdom. *Resour. Conserv. Recy.* 32: 349–358.

Cecchi, F., Pavan, P., Battistoni, P., Bolzonella, D., Innocenti, L. Characteristics of the organic fraction of municipal solid wastes in Europe for different sorting strategies and related performances of the anaerobic digestion process. *Proceedings of Latin American Workshop and Symposium on Anaerobic Digestion*, 7; Mérida, Yucatán, 22–25 Oct. 2002.

Chandler, J.A., Jewell, W.J., Gossett, J.M., Van Soest, P.J., Robertson, J.B. 1980. Predicting methane fermentation biodegradability. *Biotechnology and Bioengineering Symposium* 10: 93–107.

Chavez–Vazquez, M., Bagley, D.M., Evaluation of the performance of different anaerobic digestion technologies for solid waste treatment, *CSCE/EWRI of ASCE Environmental Engineering Conf.*, Niagara 2002 (http://gis.lrs.uoguelph.ca/agrienvarchives/bioenergy/ download/an_dig_u_toronto_2000. pdf).

Chian, E.S.K., De Walle, F.B. 1976. Sanitary landfill leachates and their treatment. *J. Environ. Eng. Div. Proc. Amer. Soc. Civ. Eng.* 102: 411–431.

Christensen, T.H. & Kjeldsen, P. 1995. Landfill emissions and environmental impact: An introduction. In Christensen, T.H., Cossu, R. & Stegmann, R., (eds.) *SARDINIA '95, Fifth International Landfill Symposium, Proceedings*, Volume 3, CISA, Cagliari, Italy.

Chynoweth, D.P., Owens, J., O'Keefe, D., Earle, J.F., Bosch, G., Legrand, R. 1992. Sequential batch anaerobic composting of the organic fraction of municipal solid waste. *Water Sci. Technol.* 25(7): 327–339.

Ciupryk, M., Gaj, K. 2004. Ekologiczne przesłanki utylizacji biogazu składowiskowego. *Energetyka i Ekologia* 2: 123–126.

Czurejno, M. 2006. Biogaz składowiskowy jako źródło alternatywnej energii. *Energetyka* 10: 777–781.

Dahlén, L., Lagerkvist, A. 2008. Methods for household waste composition studies. *Waste Manage.* 28(7): 1100–1112.

De Baere, L. 2000. Anaerobic digestion of solid waste: state of the art. *Water Sci. Tech.* 41(3): 283–290.

De Gioanni, G., Muntoni, A., Cappai, G., Milia, S. 2009. Landfill gas generation after mechanical biological treatment of municipal solid waste. Estimation of gas generation rate constants. *Waste Manage.* 29(3): 1026–1034.

den Boer, E., Jędrczak, A., Kowalski, Z., Kulczycka, J., Szpadt, R. 2010. A review of municipal solid waste composition and quantities in Poland. *Waste Manage.* 30(3): 369–77.

Directive 1999/31/EC of 26 April 1999 on the landfill of waste. Official Journal L 182, 16/07/1999 0001–0019.

Directive 2004/42/CE of the European Parliament and of the Council of 21 April 2004 on the limitation of emissions of volatile organic compounds due to the use of organic solvents in certain paints and varnishes and vehicle refinishing products EUR-Lex, European Union Publications Office. Retrieved on 2010-09-28.

Dudek, J., Klimek, P, Kołodziejak, G., Pałkowska, H., Zaleska-Bartosz, J., Buchyńska, A., Hvozdevych, O., Podolsky, M., Stefanyk, Y. Charakterystyka biogazu wytwarzanego na wytypowanych do badań składowiskach odpadów komunalnych na terenie Województwa Podkarpackiego i Obwodu Lwowsk-iego, *Prace Instytutu Nafty i Gazu* 145, Kraków 2007.

Eleazer, W.E., Odle, W.S. III, Wang, Y., Barlaz, M.A. 1997. Biodegadability of municipal waste components in laboratory-scale landfills. *Env. Sci. Tech.* 31: 911–917.

EPA 1995. Determination of landfill gas composition and pollutant emission rates at Fresh Kills Landfill, Vol. 1. *Final Project Report*, 1995.

EPA 2008. Frequently Asked Questions about Landfill Gas and How It Affects Public Health, Safety, and the Environment, *Landfill Methane Outreach Program* (LMOP) (http://www.epa.gov/lmop/docs/ faqs_about_LFG.pdf).

Erses, A.S., Onay, T.T. 2002. In situ heavy metal attenuation in landfills under methanogenic conditions. *J. Hazard. Mater.* 99: 159–163.

Fairweather, R.J., Barlaz, M.A. 1998. Hydrogen Sulfide Production during Decomposition of Landfill Inputs. *Journal of Environmental Engineering* 124(4): 353–361.

Fox, M., Noike, T. 2004. Wet oxidation pretreatment for the increase in anaerobic biodegradability of newspaper waste. *Bioresource Technol.* 91(3): 273–28.

Guidance of Landfill Gas Flaring, SEPA 2002 (www.sepa.org.uk).

Hartmann, H., Ahring, B.K. 2005. Anaerobic digestion of the organic fraction of municipal solid waste: influence of co-digestion with manure. *Water Res.* 39(8): 1543–52.

Haubrichs, R., & Widmann, R. 2006, Evaluation of aerated biofilter systems for microbial methane oxidation of poor landfill gas. *Waste Manage.* 26: 673–674.

Häusler, T., Schreier, W. 2005. Analyse siliziumorganischer verbindungen im deponiegas sowie CO-messungen zur brandfrüherkennung. Verlag Abfall aktuell – Band 16 – *Stillegung und Nachsorge von Deponien*: 241–249.

Hristovski, K., Olson, L., Hild, N., Peterson, D., Burge, S. 2007. The municipal solid waste system and solid waste characterization at the municipality of Veles, Macedonia. *Waste Manage.* 27(11): 1680–1689.

Huber-Humer, M., Gebert, J., Hilger, H. 2008. Biotic systems to mitigate landfill methane emissions. *Waste Manage. Res.* 26: 33.

Itodo, I.N., Agyo, G.E., Yusuf, P. 2007. Performance evaluation of a biogas stove for cooking in Nigeria, *Journal of Energy in Southern Africa* 18(3): 14–18.

Jaffrin, A., Bentounes, N., Joan, A.M., Makhlouf, S. 2003. Landfill biogas for heating greenhouses and providing carbon dioxide supplement for plant growth. *Biosyst Eng.* 86: 113–123.

Jędrczak, A., Pilicydis, P., *Raport z badań morfologicznych odpadów komunalnych dowożonych do składowiska w Zakurzewie*. P.B.P. EKOSYSTEM, Zielona Góra, 2000.

Jędrczak, A., Szpadt, R. 2006. *Określenie metodyki badań składu sitowego, morfologicznego i chemicznego odpadów komunalnych* (http://www.wfosigw-gda.pl/uploadfiles/raport_metody_badan_skladu.pdf z dnia 02.03.2010r.).

Jones, K.L., Grainger, J.M. 1983. The application of enzyme activity measurements to a study of factors affecting protein, starch and cellulose fermentation in domestic refuse. *Eur. J. Appl. Microbiol. Biotechnol.* 18: 181–185.

Jørgensen, P.J. *Biogas – green energy. Process, Design, Energy supply, Environment*, Faculty of Agricultural Sciences, Aarhus University 2009, 2nd edition.

Kalina, J., Skorek, J. 2002. Paliwa gazowe dla układów kogeneracyjnych, Seminarium cykliczne *Energetyka w procesie przemian* Gliwice 2002.

Kim, K.-H., Choi, Y.J., Jeon, E.C., Sunwoo, Y. 2005. Characterization of malodorous sulfur compounds in landfill gas, *Atmos. Environ.* 3(96): 1103–1112.

Kim, K-H., Baek, S.O., Choi, Y.J., Sunwoo, Y., Jeon, E.C., Hong, J.H. 2006. The emissions of major aromatic VOC as landfill gas from urban landfill sites in Korea, *Environ. Monit. Assess.* 118: 407–422.

Kjeldsen, P., Barlaz, M.A., Rooker, A.P., Baum, A., Ledin, A., Christensen, T.H. 2002. Present and long-term composition of MSW landfill leachate: *A Review. Crit. Rev. Env. Sci. Tec.* 32: 297–336.

Landrat, M., 2007. Produkcja paliw gazowych w procesach składowiskowych (Fuel gases production in landfill processes). *Przegląd Komunalny* 9: 46–48.

Lauber, J., Morris, M.E., Ulloa, P., Hasselriis, F. Comparative impacts of local waste to Energy vs. long distance Disposal of municipal waste, *Air & Waste Management Association Conference*, New Orleans, Louisiana, June 20–23, 2006 (www.energyanswers.com/pdf/awma_final.pdf, date of access Sept 2012).

Lebedev, V.S., Garbatyuk, O.V., Ivanow, D.V., Nozhevnikova, A.N., Nekrasova, V. K. 1994. Biochemical Processes of Biogas Formation and Oxidation in Municipal Waste Dump. *J. Ecol. Chem.* 3(2): 121–132.

Lee, S., Xu, Q., Booth, M., Townsend, T., Chadik, P., Bitton, G. 2006. Reduced sulfur compounds in gas from construction and demolition debris landfills. *Waste Manage.* 26: 526–533.

Lornage, R., Redon, E., Lagier, T., Hébé, I., Carré, J. 2007. Performance of a low cost MBT prior to landfilling: study of the biological treatment of size reduced MSW without mechanical sorting. *Waste Manage.* 27(12): 1755–64.

Maćków, I., Małysa, H., Sebastian, M., Szpadt, R. 2005. Zmienność składu i właściwości odpadów komunalnych miasta Wrocławia w latach 1992–2004. VI *Międzynarodowe forum gospodarki odpadami, Efektywność gospodarowania odpadami*, Poznań – Licheń Stary.

Mahar, R.B., Liu, J., Li, H., Nie, Y. 2009. Bio-pretreatment of municipal solid waste prior to landfilling and its kinetics. *Biodegradation* 20(3): 319–30.

Major, G. Learning from experiences with small-scale cogeneration. *CADDET Analyses Series* No. 1. Sitard, Netherlands 1993.

Malenshek, M., Olsen, D.B. 2009. Methane number testing of alternative gaseous fuels. *Fuel* 88: 650–656.

Mata-Alvarez, J., Llabrés, P., Cecchi, F., Pavan, P. 1992. Anaerobic digestion of the Barcelona central food market organic wastes: Experimental study. *Bioresource Technol.* 39: 39–48.

Meres, M., Szczepaniec-Cieciak, E., Sadowska, A., Piejko, K., Szafnicki, K. 2004. Operational and meteorological influences on the utilized biogas composition at the Barycz landfill site in Cracow, Poland, *Waste Manage. Res.* 22(3): 195–201.

Narayana, T., 2009. Municipal solid waste management in India: From waste disposal to recovery of resources? *Waste Manage.* 29(3): 1163–1166.

Nowakowski, S. 1997. Pozyskiwanie biogazu, *Ochrona Powietrza i Problemy Odpadów* 1.

Nyns, E.J. 1986. Biomethanation processes. In Schonborn W. (ed). *Microbial degradations*. Vol. 8. Berlin: Wiley- VCH Weinheim.

Pan, J., Voulvoulis, N. 2007. The role of mechanical and biological treatment in reducing methane emissions from landfill disposal of municipal solid waste in the United Kingdom. *J Air Waste Manag Assoc.* 57(2): 155–63.

Parker, T., Dottridge, J., Kelly, S. 2002. Investigation of the composition and emissions of trace components in landfill gas (P1–438/TR). *Technical Report. Environment Agency.* Bristol, UK.

Pawłowska, M., Czerwiński, J., Stępniewski, W. 2008. Variability of the non-methane volatile organic compounds (NMVOC) composition in biogas from sorted and unsorted landfill material. *Arch. Environ. Prot.*, 34(3): 287–298.

Persson, M., Jönsson, O., Wellinger, A. 2006. Biogas upgrading to vehicle fuel standards and grid injection, *IEA Bioenergy.*

Pipatti, R., Savolainen, I. 1996. Role of energy production in the control of greenhouse gas emissions from waste management. *Energ. Convers. Manage.* 37(6–8): 1105–1110.

Porpatham, E., Ramesh, A., Nagalingam, B. 2008. Investigation on the effect of concentration of methane in biogas when used as a fuel for a spark ignition engine. *Fuel* 87(8–9): 1651–1659.

Qamaruz-Zaman, N., Milke, M.W. 2010. Similarity of first-order rate constants for methane from food wastes in batch and continuous feed systems. Venice, Italy: *3rd International Symposium on Energy from Biomass and Waste*, 8–11 Nov.

Rao, M.S., Singh, S.P., Singh, A.K., Sodha, M.S. 2000. Bioenergy conversion studies of the organic fraction of MSW: assessment of ultimate bioenergy production potential of municipal garbage. *Appl. Energ.* 66(1) 75–87.

Rasi, S., Veijanen, A., Rintala, J. 2007. Trace compounds of biogas from different biogas production plants. *Energy* 32: 1375–1380.

Reinhart, D.R. & Townsend, T.G., 1998. Landfill bioreactor design and operation, CRC Press, New York, NY.

Roati, C., Fiore, S., Ruffino, B., Marchese, F., Novarino, D., Zanetti, M.C. 2012. Preliminary evaluation of the potential biogas production of food-processing industrial wastes. *American Journal of Environmental Sciences* 8(3): 291–296.

Sanders, W.T.M., Veeken, A.H.M., Zeman, G, van Lier, J.B. 2003. Analysis and optimisation of the anaerobic digestion of the organic fraction of municipal solid waste. In Mata-Alvarez J. (ed), *Biomethanization of the organic fraction of municipal solid wastes*, 63–89, IWA Publishing.

Scaglia, B., Confalonieri, R., D'Imporzano, G., Adani, F. 2010. Estimating biogas production of biologically treated municipal solid waste. *Bioresource Technol.* 101(3): 945–952.

Scharff, H., & Jacobs, J. 2006. Applying guidance for methane emission estimation for landfills. *Waste Manage.* 26(4): 417–29.

Scheutz, C., Bogner, J., Chanton, J.P., Blake, D., Morcet, M., Aran, C., Kjeldsen, P. 2008. Atmospheric emissions and attenuation of non-methane organic compounds in cover soils at a French landfill. *Waste Manage.* 28(10): 1892–1908.

Schievano, A., Scaglia, B., D'Imporzano, G., Malagutti, L., Gozzi, A., Adani, F. 2009. Prediction of biogas potentials using quick laboratory analyses: upgrading previous models for application to heterogeneous organic matrices. *Bioresour Technol.* 100(23): 5777–82.

Scott, P.E., Dent, C.G., Baldwin, G. 1988. The composition and environmental impact of household waste derived landfill gas, 2nd report, DOE Waste Technical Division, R&D Technical Report CWM 41, 88.

Shin, H.C., Park, J.W., Park, K., Song, H.C. 2002. Removal characteristics of trace compounds of landfill gas by activated carbon adsorption. *Environ. Pollut.* 119(2): 227–236.

Smith, L., Karim, H., Etemad, S., Pfefferle, W.C., Catalitic combustion, In: *The gas turbine handbook*, National Energy Technology Laboratory, USA, 2006.

Soltani-Ahmadi, H., 2002. *A Review of the Literature Regarding Non-Methane and Volatile Organic Compounds in Municipal Solid Waste Landfill Gas.* University of Delaware, Department of Civil and Environmental Engineering. Featured in the September/October 2002 issue of MSW Management (Forester Communications, Inc.). (www.stormcon.com/nmocvoc.pdf).

Spiegel, R.J, Thorneloe, S.A, Trocciola, J.C, Preston, J.L. 1999. Fuel cell operation on anaerobic digester gas: conceptual design and assessment. *Waste Manage.* 19(6): 389–399.

Stachowitz, W., 2001, 15 years of experience in the field of landfill gas disposal, standards, problems, solutions and procedures. In Christensen, T. H., Cossu, R. & Stegmann, R. (eds) *Proc. Sardinia 2001, Eighth International Landfill Symposium,* Margherita di Pula, Italy, Vol. II, 601—611. Published by CISA Environmental Sanitary Engineering Centre, Cagliari, Italy.

Stegmann, R. 2005. Mechanical biological pretreatment of municipal solid waste. In *Proceedings of Sardinia 2005, Tenth International Waste Management and Landfill Symposium.* CISA Ed, Italy, pp. 159–160 (book of abstract, full paper available on CD-ROM).

Tchobanoglous, G., Theisen, H., Vigil, S., 1993. Integrated solid waste management, engineering principles and management issues, McGraw-Hill, New York.

US EPA. 2003. *characterization of municipal solid waste in the United States*: 2001 Update, EPA 530-R-03-011, Washington, DC.

Wheless E., Pierce J. Siloxanes in Landfill and Digester Gas Update. SWANA 27th LFG Conference. March 22–25, 2004.

Williams T. 2009. European gas interchangeability, *4th World Gas Conference.* 5–9 October 2009, Buenos Aires, Argentine.

Woelders, H., Moorman, F.J.A., Van der Ven, B.L., Glas, H., Coops, O. 1993. Landfilling of MSW after separation of biowaste and RDF: Emission control. In *Proceedings Sardinia 93 - 4th International landfill Symposium,* CISA publisher, Cagliari, Italy, vol. II. 2057–2070.

Zamorano, M., Perez, J.I., Paves, I.A., Ridao, A.R. 2007. Study of the energy potential of the biogas produced by an urban waste landfill in Southern Spain. *Renew. Sust. Energ. Rev.* 11: 909–922.

Zhu, B, Gikas, P, Zhang, R, Lord, J., Jenkins, B., Li, X. 2009. Characteristics and biogas production potential of municipal solid wastes pretreated with a rotary drum reactor. *Bioresource Technol.* 100: 1122–1129.

Zhuang, Y., Wu, S.W., Wang, Y.L., Wu, W.X., Chen, Y.X. 2008. Source separation of household waste: A case study in China. *Waste Manage.* 28(10): 2022–2030.

Zou, S.C., Lee, S.C., Chan, C.Y., Ho, K.F., Wang, X.M., Chan, L.Y., Zhang, Z.X. 2003. Characterization of ambient volatile organic compounds at a landfill site in Guangzhou, SouthChina. *Chemosphere* 51(9): 1015–1022.

CHAPTER 3

Increasing landfill gas production and recovery

3.1 INTRODUCTION

An increase in the anaerobic decomposition rate of waste leads to a rise in landfill gas production and faster stabilization of waste. This goal can be achieved by converting a conventional landfill into an anaerobic bioreactor landfill. Landfill Gas (LFG), withdrawn in a controlled manner, is used as an energy source. Consequently, carbon dioxide having a significantly lower global warming potential than methane (ca. 21 times) is released into the atmosphere. An anaerobic bioreactor landfill (ABL) is a variant of the bioreactor landfill concept, which assumes optimization of the operational parameters and environmental conditions within the waste mass in order to obtain a "stable waste" within a reasonable time scale, thus ensuring limited risk to the environment, even when liner failure occurs (Westakle, 1997).

In practice, ensuring optimal conditions inside the bioreactor landfill is based on the control of waste moisture. The term "control" is to be understood as not only maintaining optimum moisture content, but also the movement of water in the waste layer (Warith et al. 2005). Landfill leachate is the most common source of water used for waste wetting. It is recycled into the interior of the landfill (Fig. 3.1). Sometimes, however, its volume is too small to ensure sufficient moistening of waste and it is necessary to include an additional water source (e.g. municipal sewage sludge or industrial wastewater). A landfill, inside which recirculation is practiced only as a means of disposing of leachate is not considered to be a bioreactor (Townsend et al. 2008). The legal definition adopted in the United States (USEPA Clean Air Act, 40 CFR 63.1990, National Emissions Standards for Hazardous Air Pollutants) constrains the term "bioreactor landfill", emphasizing the presence of an additional water source. It defines a bioreactor landfill as "a MSW landfill or a portion of a MSW landfill where any liquid, other than leachate (leachate includes landfill gas condensate) is added in a controlled fashion into the waste mass (often in combination with recirculating leachate) to reach a minimum average moisture content of at least 40% by weight to accelerate or enhance the anaerobic (without oxygen) biodegradation of the waste".

The concept of an "anaerobic bioreactor landfill" has emerged as an alternative to a "dry tomb" landfill, the operation of which is based on isolating the waste mass from the environment. Noticeably, the waste biodegradation process that occurs in them is very slow. Thus it is necessary to ensure long-term control of such a facility. Waste humidity which is insufficient for the growth of microorganisms responsible for anaerobic digestion is the main reason for the slowdown in the mineralization rate of organic matter. Recognizing these limitations, efforts were focused on improving water conditions within the landfill. Research on faster stabilization of landfilled waste was initiated in the United States in the 1970s and to date the country is a leader in the application of "bioreactor landfill" technology.

A properly operated bioreactor landfill can provide a more environmentally friendly waste management strategy compared to a conventional landfill. Control of gas and leachate collection, which is the basis for this technology, contributes to the improvement of air, water and soil quality. Enhancing the microbiological processes in the waste by providing unlimited water accessibility accelerates the decomposition process of readily and moderately biodegradable organic components, which shortens the waste stabilization time to 5–10 years in comparison to over 30 years in a conventional landfill (Pacey et al. 1996). In the context of extensively reducing the landfill lifetime, ABL technology should be considered as a sustainable landfill method. It makes it possible to limit the period the landfill impacts on the environment to the lifetime of one generation. The application of this technology supports

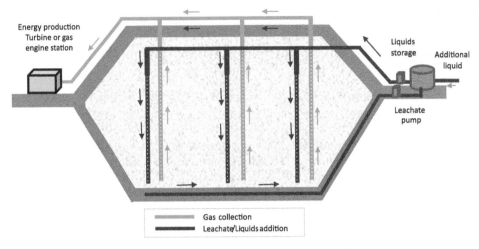

Figure 3.1 Concept of the anaerobic bioreactor landfill (according to US EPA).

an increase in the consumption of an alternative energy source instead of fossil fuels. Other benefits are related to more efficient use of landfill capacity and minimizing leachate treatment and storage problems. Technically, an anaerobic bioreactor landfill does not require any substantive changes in cell construction and equipment when compared to a sanitary landfill. It is equipped with liners, leachate collection and gas management systems. The main modifications are related to the leachate management system, which has to ensure an efficient and continuous recirculation of leachate.

3.2 RECIRCULATION OF LIQUIDS AS A BASIS FOR AN ANAEROBIC BIOREACTOR LANDFILL

3.2.1 *Increasing moisture content and water migration inside deposited waste*

The fundamental rule of anaerobic landfill bioreactor technology is to ensure sufficient amounts of water for the microorganisms responsible for the anaerobic digestion of waste, throughout the whole of their biochemical activity. The activity of fermentative bacteria, and thus the rate of biodegradation of organic matter and biogas production increases with the water content in the substrate. This is confirmed by studies conducted on various types of waste. Methane yield obtained during digestion of the organic fraction from municipal solid waste was 17% lower at 70% moisture content by weight than at 80% (Fernandez et al. 2008). In addition, methane production during digestion of food waste was higher at 80% waste moisture content than at 75% and 70% (Forster-Carneiro, 2008). The limiting influence of moisture was also observed in the case of such wet substrates as dewatered sewage sludge (with water content ranging between 89–97%). Biogas production under mesophilic conditions decreased when the moisture content of sludge fell below 91.1% (Fujishima et al. 2000).

Taking into account the predominant role of water during anaerobic digestion, it is assumed that the moisture content in the substrates subjected to methane fermentation, targeted at biogas production, should not be less than 60% by weight. In solid-state anaerobic digestion (dry fermentation) technologies, which are widely used in the municipal solid waste digestion process the preferred moisture content is in the range of 70–80% by weight (Oleszkiewicz & Poggi-Varaldo, 1997). While, the average moisture content of MSW deposited in landfills is lower, it varies over a wide range between 15–65% by weight (Tchobanglous, 1993, Chandrappa & Das, 2012, Das & Bhattacharyy, 2013). The wettest component of the municipal waste stream, with a moisture content reaching up to 80%, is food waste, but it usually constitutes only a small percentage of the total waste mass. For example food waste constituted 14.5% of the total MSW generated in 2011 in the United States (U.S. EPA, MSW data). The primary amount of water contained in the waste ranges between 15–40% and does not fully cover the needs of the microorganisms. Therefore it is necessary to supply additional water. The results from laboratory and field studies show that leachate recirculation alone does not ensure optimal process conditions. Only when a supplementary water source, such as sewage sludge, industrial wastewater, clean water, storm water runoff, or sewage effluents (Brochure; Gupta et al. 2007) is implemented, suitable moisture content may be reached.

From a practical point of view, water field capacity and hydraulic conductivity are very important physical waste parameters. These properties determine the amount of leachate and their formation rate. Based on the values of these parameters the operational conditions for liquids recirculating inside the landfill can be determined. These parameters depend mainly on waste particle size and organic matter content, which determine pore size distribution and the porosity of the medium. Field capacity is the value of water content held in the material after the gravitational water has drained away. The field capacity of municipal solid waste varies over a wide range between 14–44% v/v (Zeiss & Major, 1992/93, Bengtsson et al. 1994). According to an evaluation by Reinhardt and Ham (1974) the volume of liquid addition, which allows a water field capacity in waste to be reached varies between 0.1–0.2 m^3 per Mg of solid waste (25,000–50,000 gallons per 1,000 tons).

The rate of organic matter biodegradation is affected not only by an increase in waste humidity but also by the movement of liquid in the deposited waste profile which is important. It promotes an increase in the biodegradation rate, contributing to a uniform distribution of microorganisms, nutrients and heat. This provides better contact between the microorganisms and substrates, and dilutes toxins that may have accumulated locally in too high concentrations. Knowledge of hydraulic conductivity values allows for the maintenance of proper water movement in the waste layer and prevents water stagnation on its surface. Hydraulic conductivity is a parameter that describes water movement through the porous media. It depends on the physical properties of the material but also on the degree of pore saturation by a liquid phase. When the pores are completely filled with water the saturated hydraulic conductivity (Ks) is determined. The permeability of the landfill body is a function of several parameters including waste composition, form of compaction, homogeneity,

Table 3.1   Moisture content [% of weight] of municipal solid waste.

Waste category	Chandrappa & Das (2012)	Valkenburg et al. (2008)	Gidarakos et al. (2005)
MSW	15–40	21.6	36.72
Cardboard, corrugated paper, paper	4–10	5.2	17.11
Food waste	50–80	36.6	66.99[4]
Garden waste	30–80	43.4[1]	
		31.0[2]	
Glass	4	–	–
Wood	15–40	46.7[3]	21.41
Textiles	6–15	13.5	
Leather	8–12	–	
Rubber	1–4		
Plastic	1–4	0.2	2.93
Metal	2–6	–	

[1] leaves and grass; [2] prunings and trimmings; [3] branches and stump; [4] putrescible (food waste, yard waste, leaves).

landfill height and the build-up of the landfill layers (Mathlener et al. 2006). According to the results of a study carried out by Staub et al. (2009) and the literature reviews that have been carried out by them, the saturated hydraulic conductivity of waste ranged from 5.7×10^{-8} to 1.0×10^{-5} m s^{-1} in the field tests, and from 1.0×10^{-8} to 1.0×10^{-4} m s^{-1} in the laboratory studies. The saturated hydraulic conductivities of municipal solid waste measured at a landfill in Florida, using the borehole permeameter tests were between 5.4×10^{-8} m s^{-1} and 6.1×10^{-7} m s^{-1} and also fell inside the mentioned range (Jain et al. 2006). The value of hydraulic conductivity decreased with increasing waste density (Staub, 2009). The Ks values for waste were in a range of the values measured in sand under field conditions—which were ca. 8.90×10^{-5} m s^{-1} (Widomski et al. 2012), and in the samples taken from the surface layer of Alfisols measured in laboratory test (in the Wit apparatus)—which ranged from 1.29×10^{-6} to 4.48×10^{-6} m s^{-1} (Iwanek, 2008).

3.2.2   *Other effects accompanying the supply of liquids to a landfill bioreactor*

Many different compounds that can change the chemical conditions inside the waste mass are introduced along with the irrigating liquids into the landfill. Some of them may have a direct effect on the production of biogas (the substrate for microorganisms) while others may influence indirectly, affecting the growth of microorganisms or modifying chemical transformations in the landfill.

One of the beneficial effects of leachate recirculation is related to its high alkalinity. Alkalinity acts as a buffer for the acidic products of organic matter decomposition, and thus accelerates the start of methanogenesis, enabling it to reach the optimum pH value (6.8–7.4) for the growth of methanogens. Therefore, a very important feature of the leachate is the age of the landfill from which it is derived as its alkalinity increases with age. According to Lebiocka (2013) the alkalinity of the leachate taken out from a middle-aged landfill (5–10 years) was 8.500 ± 187 mg dm^{-3}, while from an old landfill (over 10 years) it was 15.050 ± 658 mg dm^{-3}. A favourable effect resulting from a mature landfill leachate supply to a young landfill was confirmed by Erses & Onay (2003) in a laboratory experiment. An increase in methanogenic population activity and an acceleration of waste stabilization was observed. These effects can be attributed mainly to the high buffering capacity of the leachate and waste inoculation with methanogens. Similar benefits were observed when the anaerobically digested sewage sludge was introduced into the landfill together with recirculated leachate (Knox, 1997). Municipal solid waste and leachate are the media, so rich in different chemicals that additional nutrients supplied to the leachate stream recirculating through the waste does not seem to provide any further enhancement to the process (Tittlebaum, 1982).

3.3   TECHNICAL REQUIREMENTS FOR ANAEROBIC BIOREACTOR LANDFILL CONSTRUCTION

An anaerobic bioreactor landfill utilises the same organic matter degradation processes to that which occur in a conventional sanitary landfill. The differences are mainly technological modifications, which change the conditions inside the landfill. These modifications aim to enhance the microbiological processes which transform and stabilize the readily and moderately decomposable organic waste constituents within 5–10 years of implementing the bioreactor processes (Pacey et al. 1996). Quick-filling the landfill cell with organic-rich MSW, its completion with impermeable gas tight capping and leachate recirculation in the waste mass are the means by which microbiological processes are enhanced (Jones et al. 2007). However, the bioreactor processes require significant additional liquid to reach and maintain optimal conditions. Leachate alone is usually not available in sufficient quantities to sustain the bioreactor processes. Water, other non-toxic or non-hazardous liquids, and semi-liquids are suitable alternatives as supplementary leachates (depending on climatic conditions and regulatory approval (Pacey et al. 1996).

From the technical perspective, the construction of a bioreactor landfill is similar to a conventional sanitary landfill. The required design components include a base liner, leach-ate collection facilities, gas collection and management facilities, and the final cap. The most significant modifications are to the leachate management system. In a traditional landfill the collected leachate is treated in a special treatment plant or is pre-treated prior to entering the municipal wastewater treatment plant. In ABL technology leachate is returned to the waste mass. Efficient systems for the collection, storage, transportation, and recycling of the leach-ate are necessary to ensure a successful operation of the bioreactor landfill.

The key issues that must be taken into account when designing a bioreactor are as follows:

- For economic and regulatory reasons a bioreactor is built as a deep cell which should be completed within 2–5 years (Gupta et al. 2007). But, too deep a cell could cause a problem due to the slow infiltration of liquids through the heavily compressed waste layer in the lower part of the bioreactor landfill.
- Providing an additional source of liquid when the amount of leachate is insufficient to ensure water content is over 60% and the development of the reactor's management system.
- Designing a system for leachate recirculation and distribution of liquids inside the bioreac-tor landfill.

The possible solutions for the lead-in of liquids are:

- to add the liquids to the waste before landfilling (pre-wetting); this is simple, efficient and ensures uniformity but is labour-intensive, incompatible with closure (FBDP, 2008).
- to add the liquids at the working face while the waste is being placed into the cell (dur-ing active landfilling); operators must be prepared for a rapid increase in gas production (Pacey et al. 1996).
- to add the liquids after the all the waste mass is placed inside the cell, and capping has been completed; this allows to control the intensity of gas production (Pacey et al. 1996).

Adding liquids to waste can be accomplished by using different techniques described by Townsend (2003):

Surface leachate recirculation (pre-cap) methods:

○ spray irrigation,
○ drip irrigation,
○ tanker truck application,
○ infiltration ponds.

Surface leachate recirculation (post-cap) methods:

○ leach field,
○ surface trenches,
○ drip irrigation.

Sub-surface leachate recirculation methods:

○ vertical injection wells,
○ horizontal trenches,
○ buried infiltration galleries.

Climate conditions, environmental impacts, availability of liquids, financial means, and legal regulations are the main criteria when selecting the liquid recirculation method. The two most preferred options from the standpoint of environmental impact are buried trenches and vertical wells. They offer minimum exposure pathways, good all-weather performance, and favourable aesthetics, but they may be sensitive to the impact of differential settlement (Pacey et al. 1996). If the leachate is planned to be recycled by direct spray irrigation onto landfill surfaces covered with vegetation it would be beneficial to aerate the leachate prior to recirculation. Such pre-treatment of the leachate leads to its partial mineralization which supports vegetation growth by providing nutrients (Robinson et al. 1982).

- Adapting the mechanical strength parameters of the landfill's base and side walls, as well as those of the LFG collection and leachate drainage systems, to the conditions arising as a result of the increased weight of waste. The addition of liquids to solid waste increases its specific weight. It is estimated that waste deposited in a landfill may be up to 30% heavier because of moisture uptake and settlement (Pacey et al. 1996).
- Providing very efficient gas collection and recovery systems capable of handling intensive biogas production. Horizontal trenches, vertical wells, near surface collectors, and hybrid systems can be used for gas extraction. Larger gas collection pipes, more efficient blowers and related equipment should be used where necessary. In the case of an insufficiently efficient degassing system gas pressure build-up inside the landfill may pose a threat to the landfill cover, creating conditions for geomembrane ballooning and leading to soil layer destruction (Gupta at al. 2007).
- Consideration of rapid waste settlement during the design of the landfill capping system. Accelerated settlement could be very beneficial when overfilling refuse above the design grade before placing the final cover (Gupta et al. 2007).
- More detailed control and monitoring of waste settlement, landfill gas, and leachate production and composition during the bioreactor's operational phase is necessary.
- Providing high organic matter content and a high specific surface area in the waste by mechanical pre-treatment (waste segregation, shredding).

3.4   EFFECTS OF LIQUIDS RECIRCULATING INSIDE THE LANDFILL

Water addition and its downward migration through the waste layer results in:

- Increased LFG production. High calorific LFG is generated earlier in the bioreactor landfill and at a higher rate than in a conventional landfill. The apparent first-order rate constant for methane production, measured in a demonstration cell at Central Landfill in Yolo County (California) was greater than $k = 0.4$ per year. It was more than ten-fold the typical value for a waste mass of this size (Augenstein et al. 2001). The highest rate of LFG production was observed shortly following the closure of the landfill. The rate quickly declined over the next 5 to 10 years to a relatively low but stable level (Pacey et al. 1996). The model study carried out by Jones et al. (2007) showed that the gas yield peak was significantly higher in the bioreactor then in a conventional landfill. The bioreactor reduced the gas curve's tail-off time and consequently shortened the active gas management period (Jones et al. 2007). A laboratory study into the anaerobic digestion of domestic solid waste carried out by Sponza & Ăgdăg (2004) showed a higher methane content in the biogas produced in the reactor in which leachate recirculation was applied. Although the total time of LFG production in the ABL is shorter compared to a conventional landfill, the unit gas yield and total gas yield are higher. Consequently, the recoverable gas yield from an anaerobic bioreactor is consistently higher than that produced by the "dry tomb" landfill. Increased LFG production has both economic and environmental benefits. It leads to an improvement of economics for LFG recovery and shortens the time in which lean LFG is produced, contributing to a reduction in fugitive gas emissions.
- Shortening the waste stabilization time. The decomposition of waste in an anaerobic bioreactor landfill lasts years *versus* decades in "dry tombs". Ensuring an adequate amount of water accelerates the decomposition of organic matter. According to Pacey et al. (1996) almost all of the rapid and moderately biodegradable organic constituents are decomposed

within 5–10 years of implementing a bioreactor. Faster decomposition of organic matter leads to a shortening of the period in which the landfill poses a threat to the environment. This is extremely important because of the risk of long-term instability of landfill liners. Due to faster exhaustion of organic matter resources which are susceptible to biodegradation, the landfill monitoring time carried out under post-closure care can be reduced.

- Rapid settlement of waste mass. The waste layer, through which leachate is recirculated settled faster than where there is no recirculation. Efficient settlement of waste in an ABL is caused by an increase in the waste mass due to moistening and by an increase in the rate of organic matter decomposition. It is estimated that 15–30% of landfill space is gained due to an increase in the density of waste mass caused by liquids recirculating (Benson et al. 2007). The obtained space enables the amount of waste that can be placed into the permitted landfill airspace to be increased (Pacey et al. 1996). It is a measurable, economic benefit for landfill operators. Rapid settlement was observed both in the laboratory and in the field scale. For example, leachate recirculation at the Trail Road Landfill (Canada) site during the active filling phase enhanced waste settlement, which resulted in the recovery of 25% of the landfill cell volume (Warith, 2002). Yazdani et al. (2011) noticed that settlement in large-scale cells with liquids recirculation was accelerated two to three-fold, compared to a dry-tomb pilot-scale control cell at the same point in time. The settlement rate at the Salem County Landfill (New Jersey) site has been observed at about 1.5 metres per year greater than the previously observed settlement rate before leachate recirculation commenced (US EPA, 2007).
- Decrease of pollutant concentrations in the leachate. Positive effects in the reduction of leachate strength in a full-scale bioreactor landfill were observed at Sea Carr Landfill. The COD in the leachate from the area with recirculation was always lower than in the control area (without recirculation). The differences became more apparent with time. After three years of observation the COD values in the recirculated leachate were approximately 15 g dm^{-3}, while in the control they exceeded 50 g dm^{-3}. A similar trend was observed in the case of chlorine, and ammoniacal-N. During three years of observations concentrations of these substances decreased in both cells, but the lower values and a more pronounced drop were noticed in the cell where leachate recirculation was applied. After three years, the chlorine and ammoniacal-N concentrations in the recirculated leachate was about 1,600 and 500 mg dm^{-3} respectively, while in the leachate from the control cell they were about 2,100 and 800 mg dm^{-3}, respectively (Bioreactor Landfill for Sustainable..., 2004). Recirculation of leachate from an active or closed landfill cell can be considered as an *in situ* leachate treatment method. It minimizes the hazard to the environment and reduces the leachate treatment costs (Reinhart et al. 2002). Due to the continuous recirculation of leachate the operator does not bear the costs associated with its pre-treatment and transport to the treatment plant.

3.5   CRITICAL APPROACH TO ANAEROBIC BIOREACTORS LANDFILL TECHNOLOGY

Liquid recirculation in a landfill is linked to the risk of adverse effects associated with intensive gas production and fast settlement of the waste mass. The main potential disadvantages are:

- Increased risk of fires or explosion. Rapid degradation of organic matter is a reason for the fast production of LFG. Gas venting systems used in conventional landfills may not be able to efficiently vent the bioreactor landfills which could cause an explosion (US EPA, 2006).
- Increased risk of damage to the base liner, the gas and leachate drainage systems due to the higher density of waste.
- Risk of damage to the capping system due to intensive gas production that can cause geomembrane ballooning or a rapid collapse of the waste layer.

- Risk of surface seepages. The low permeability of the material that is used as a daily or intermediate cover for the landfilled waste can cause lateral migration of the liquids. Soh & Hettiaratchi (2009) showed that when silt loam was used as a cover, less than 1% of the leachate seeped through the cover layer, but when sand was used no lateral leachate migration was observed and 100% of the leachate seeped through the cover.
- Risk of chloride and ammonia nitrogen accumulation in landfills, which could inhibit methanogenesis. To eliminate the problem of ammonia's toxic effects, leachate should be pre-aerated before recirculation (Francois et al. 2007).
- Intensified odour nuisance. Compared to traditional landfills, bioreactor landfills produce more gases, such as H_2S, which have an extremely putrid smell.
- Increased sanitary risk due to surface leachate recirculation.
- Higher initial capital costs of landfill cell construction (e.g. double-lined containment systems) and infrastructure (liquid recirculation system).
- Increased requirement for monitoring and control during the operational phase of the landfill's life.

3.6 HYBRID BIOREACTOR LANDFILL

Another type of bioreactor landfill, in which production of high calorific LFG occurs is the hybrid (aerobic-anaerobic) bioreactor landfill. In this solution waste degradation is enhanced by employing a sequential aerobic-anaerobic treatment (Technical/Regulatory Guideline, 2006). In a hybrid bioreactor landfill the waste is first degraded under aerobic and then under anaerobic conditions. Air can be injected using horizontal wells or by vacuum induction. The presence of air prolongs the initial aerobic phase of waste decomposition. The oxygen increases the waste degradation rate and the organic matter's susceptibility for biodegradation. The products from the aerobic decomposition of organic matter are metabolized by methanogens in the lower, anaerobic part of the landfill. Consequently, energy recovery from LFG starts earlier than in a traditional landfill (Rich et al. 2008). The gas collection/extraction system is not used during air injection. When filling the cell, the degradation constant is low. It increases after the cell has been filled and leachate is being recirculated.

In addition to accelerated waste decomposition, the other important advantage of hybrid bioreactor landfill technology is the high efficiency of nitrogen removal from the leachate. Oxy-

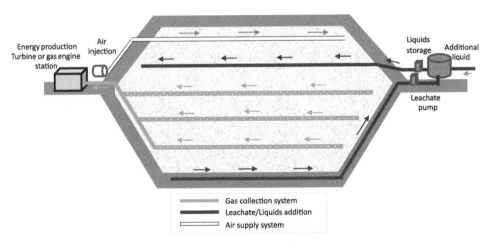

Figure 3.2 Concept of hybrid (aerobic-anaerobic) bioreactor landfill (according to US EPA).

gen injected into the upper part of the landfill modifies the redox conditions in a vertical profile through the waste layer, leading to the creation of three zones: aerobic, anoxic and anaerobic. This variation influences the vertical diversity of microbial consortia. Research carried out by He & Shen, (2006) on a laboratory scale into a hybrid bioreactor landfill system, in which air was supplied by using intermittent aeration at the top of the landfilled waste, showed multiply populations of nitrifying bacteria in the aerated waste layer and denitrifying bacteria in the middle layer of the waste. The denitrifying bacteria population was between 4 and 13 orders of magnitude greater when compared with conventional anaerobic landfilled waste layers. Due to the activity of these microorganisms very low concentrations of NH_4^+–N and TN were observed in the leachate after 105 days of bioreactor operation (186 and 289 mg dm^{-3} respectively). These parameters were still above 1,000 mg dm^{-3} in the leachate produced in the bioreactor without aeration. The concentration of NO_3^-–N was also very low (He & Shen, 2006).

The aerobic-anaerobic bioreactor landfill technology is not widespread in the world. It was applied in the field scale in North America at the Metro Recycling/Disposal landfill in Wisconsin (Kidman et al. 2011).

3.7 SUMMARY

Anaerobic bioreactor landfill technology is an efficient way of reducing gas emissions from the waste management sector. The more controlled conditions of the process versus the conventional landfill make it possible to capture greater amounts of LFG for its utilization. Such a landfill may be considered as a large-scale digester for MSW conversion to energy. Use of LFG for energy generation leads to a reduction in fossil fuel consumption. Intensifying anaerobic digestion of waste shortens the period in which lean LFG (with low methane content) is produced. In conventional landfills lean gas is usually released into the atmosphere, because of technical problems with its combustion. Thus, efficient gas capture and recovery results in a noticeable abatement of greenhouse gas emissions.

ABL technology can be regarded as a safe method of waste disposal. No fires and damage to the landfill infrastructure have been reported during the many years of observations on pilot field-scale landfills (EPA 2007). Therefore, ABL technology can be considered as an alternative to reducing landfilling of biodegradable municipal waste as recommended in the European Union.

REFERENCES

Augenstein, D., Yazdani, R., Kieffer, J., Barlaz, M. & Benemann, J. 2001. Yolo County Controlled Landfill Project. *First National Conference on Carbon Sequestration*, May 15–17, Washington, DC.

Bengtsson, L., Bendz, D., Hogland, W., Rosqvist H. & Åkesson M. 1994. Water balance for landfills of different age. *Journal of Hydrology* 158 (3–4): 203–217.

Benson, C.H., Barlaz, M.A., Lane, D.T. & Rawe, J.M. 2007. Practice review of five bioreactor/recirculation landfills. *Waste Manag.* 27(1): 13–29.

Bioreactor landfill for sustainable solid waste landfill management. 2004. Kasetsart University, Thailand (http://www.swlf.ait.ac.th/data/pdfs/BLSS1.pdf).

Brochure Waste Management Bioreactor Program (http://www.wm.com/sustainability/pdfs/bioreactor-brochure.pdf).

Chandrappa, R. & Das, D.B. 2012. *Solid Waste Management*, Environmental Science and Engineering, Springer-Verlag Berlin Heidelberg.

Das, S. & Bhattacharyy, B.K. 2013. Municipal solid waste characteristics and management in Kolkata, India. *International Journal of Emerging Technology and Advanced Engineering* 3(2): 147–152.

Erses, S.A. & Onay, T.T., 2003. Accelerated landfill waste decomposition by external leachate recirculation from an old landfill cell. *Water Sci Technol.* 47(12):215–22.

FBDP, 2008. *Florida Bioreactor Demonstration Project.* Final Report (http://www.bioreactorlandfill.org/BioreactorFinalReport/TableofContents.htm)

Fernandez, J., Perez M. & Romero L.I. 2008. Effect of substrate concentration on dry mesophilic anaerobic digestion of organic fraction of municipal solid waste (OFMSW). *Bioresource Technology* 99(14): 6075–80.

Forster-Carneiro, T., Pérez, M. & Romero L.I. 2008. Influence of total solid and inoculum contents on performance of anaerobic reactors treating food waste. *Bioresource Technology* 99(15): 6994–7002.

Francois, V., Feuillade, G., Matejka, G., Lagier, T. & Skhiri, N. 2007. Leachate recirculation effects on waste degradation: Study on columns. *Waste Management* 27: 1259–1272.

Fujishima, S., Miyahara, T. & Noike, T. 2000. Effect of moisture content on anaerobic digestion of dewatered sludge: ammonia inhibition to carbohydrate removal and methane production. *Water Sci Technol.* 41(3): 119–27.

Gidarakos, E., Havas, G. & Ntzamilis, P. 2006. Municipal solid waste composition determination supporting the integrated solid waste management system in the island of Crete, *Waste Management* 26(6): 668–79.

Gupta, S., Choudhary, N. & Alappat, B.J. 2007. Bioreactor Landfill for MSW Disposal in Delhi *Proceedings of the International Conference on Sustainable Solid Waste Management*, 5–7 September 2007, Chennai, India: 474–481.

He, R. & Shen D.-S. 2006. Nitrogen removal in the bioreactor landfill system with intermittent aeration at the top of landfilled waste. *Journal of Hazardous Materials* 136(3): 784–790.

Iwanek, M. 2008. A method for measuring saturated hydraulic conductivity in anisotropic soils. *Soil Science Society of America Journal* 72(6): 1527–1531.

Jain, P., Powell, J., Townsend, T. & Reinhart, D. 2006. Estimating the hydraulic conductivity of landfilled municipal solid waste using the borehole permeameter test. *J. Environ. Eng.* 132(6): 645–652.

Jones, T., Gregory, R. & Wilson, A. 2007. *Role of landfill gas utilization in sustainable landfilling,* Report for Sustainable Landfill Foundation.

Kidman T., Beddingfield, E., Minnucci C., Cook, M. & Van Abe, C. 2011. *Aerobic and Anaerobic Bioreactor Project Protocol.* Prepared by Science Applications International Corporation for the Climate Action Reserve. December 19, 2011.

Knox, K. 1997. *A review of the Brogborough and landfill 2000 test cells monitoring data.* Final Report for the Environment Agency, Report No. CMW 145/97. (Research Contract No. EPG 1/7/11). Knox Associates, Nottingham.

Lebiocka, M. *Utylizacja odcieków składowiskowych na drodze współfermentacji z osadami ściekowymi* (Landfill leachate utilization by co-digestion with municipal sewage sludge), Ph.D thesis. Lublin 2013.

Mathlener, A., T. Heimovaara H. Oonk L. Luning H.A. van der Sloot A. van Zomeren R. 2006. *Opening the black box, process-based design criteria to eliminate aftercare of landfills.* Dutch Sustainable Landfill Foundation.

Oleszkiewicz, J.A. & Poggi-Varaldo H. 1997. High-solids anaerobic digestion of mixed municipal and industrial wastes. *J. Environ. Eng.* 123: 1087–1092.

Pacey, J., Augenstein, D., Morck, R., Reinhart, D. & Yazdani, R. 1996. *The bioreactor landfill - an innovation in solid waste management* (http://www.epa.gov/projctxl/yolo/tech5.pdf)

Reinhart, D., McCreanor, P. & Townsend, T. 2002. The bioreactor landfill: Its status and future. *Waste Management and Research* 20(2): 162–171.

Reinhart, J.J. & Ham R.K. 1974. *Solid waste milling and disposal on land without cover.* U.S. Environmental Protection Agency, Cincinnati, Ohio, PB-234 930.

Rich, C., Gronow, J. & Voulvoulis N. 2008. The potential for aeration of MSW landfills to accelerate completion. *Waste Management* 28: 1039–1048.

Robinson, H.D., Barber, C. & Maris, P. J. (1982). Generation and treatment of leachate from domestic wastes in landfills. *Water Pollution Control* 81: 465

Soh, I. & Hettiaratchi, J. 2009. Potential lateral migration of leachate in flushing bioreactor landfills during aggressive leachate recirculation. Pract. Period. Hazard. Toxic Radioact. *Waste Manage.* 13. SPECIAL ISSUE: Advances in Solid Waste Management Technology: 174–178.

Sponza, D.T. & Ăgdăg, O.N. 2004. Impact of leachate recirculation and recirculation volume on stabilization of municipal solid wastes in simulated anaerobic bioreactors. *Process Biochemistry* 39: 2157–2165.

Staub, M., Galietti, B., Oxarango, L., Khired, M.V. & Gourc, J.-P. 2009. Porosity and hydraulic conductivity of MSW using laboratory-scale tests. *Proceedings of the Third International Workshop "Hydro-Physico-Mechanics of Landfills"* 10–13 March. Braunschweig, Germany.

Tchobanoglous, G., Theisen, H. & Vigil, S. 1993. Integrated Solid Waste Management, Engineering Principles and Management Issues, McGraw-Hill, New York.

Technical and regulatory guideline for characterization, design, construction, and monitoring of bioreactor landfills (ALT-3, 2006), Prepared by Interstate Technology & Regulatory Council, Washington, DC (www.itrcweb.org/Guidance/GetDocument?documentID = 3).

Tittlebaum, M.E. 1982. Organic carbon content stabilization through landfill leachate recirculation. *Journal of the Water Pollution Control Federation* 54(8): 428–433.

Townsend, T. 2003. Liquid systems and monitoring. *Bioreactor Landfill Workshop*. February 27–28. Arlington, VA.

Townsend, T., Kumar, D. & Ko, J. 2008. *Bioreactor Landfill Operation*: A guide for development, implementation and monitoring: version 1.0 (July 1, 2008). Prepared for the Hinkley Center for Solid and Hazardous Waste Management, Gainesville, FL.

U.S. EPA 2007. *Bioreactor Performance Summary Paper*, EPA530-R-07-007 U.S. EPA Office of Solid Waste, Municipal and Industrial Solid Waste Management Division.

U.S. EPA website (http://www.epa.gov/osw/nonhaz/municipal/landfill/bioreactors.htm#5).

U.S. EPA, MSW data (http://www.epa.gov/epawaste/nonhaz/municipal/index.htm).

U.S. EPA. 2006. Wastes - Non-Hazardous Waste - Municipal Solid Waste: Bioreactors. Retrieved February 3, 2010, (http://www.epa.gov/waste/nonhaz/municipal/landfill/bioreactors.htm).

USEPA Clean Air Act, 40 CFR 63.1990, National Emissions Standards for Hazardous Air Pollutants.

Valkenburg, C. Gerber, M.A. Walton, C.W. Jones, S.B. Thompson, B.L. & Stevens, D.J. 2008. Municipal solid waste (MSW) to liquid fuels synthesis. Volume 1: *Availability of Feedstock and Technology.* Pacific Northwest National Laboratory Richland, Washington 99352.

Warith M., Li X. & Jin H. 2005. Bioreactor landfills: state-of-the-art review. *Emirates Journal for Engineering Research* 10(1): 1–14.

Warith M. 2002. Bioreactor landfills: experimental and field results. *Waste Management* 22(1): 7–17.

Westakle, K. 1997. Sustainable landfill – possibility or pipe dream, *Waste Management and Research* 15: 453–461.

Widomski M.K., Iwanek M. & Stepniewski W. 2012. Implementing anisotropy ratio to modeling of water flow in layered soil. *Soil Science Society of America Journal* 77(1): 8–18.

Yazdani, R., Augenstein, D., Kieffer, J., Sananikone, K. & Akau, H. 2011. Accelerated anaerobic composting for energy generation at Yolo County Central Landfill. California Energy Commission, *PIER Renewable Energy Technologies Program*.

Zeiss, C. & Major, W. 1992/93. Moisture flow through municipal solid waste: patterns and characteristics. *Journal of Environmental Systems* 22 (3): 211–223.

CHAPTER 4

Attenuation of greenhouse gas emissions *via* landfill aeration

4.1 INTRODUCTION

One of the methods for reducing greenhouse gas emissions from landfills is to aerate the deposited waste. The decomposition of organic matter occurring in the presence of oxygen is faster and more efficient than anaerobic digestion. The transformation of the conditions within the waste body from anaerobic to aerobic leads to a decrease in a methane concentration in the landfill gas in favor of carbon dioxide (having significantly lower GWP). What is more, the presence of oxygen provides other benefits. For instance, it also contributes to the reduction of leachate pollution and rapid waste settlement. Residual waste consists of hardly and non-biodegradable compounds that do not pose any threat to the environment.

The effectiveness of landfills aeration has been tested in various countries around the world; both in Europe (Germany, Austria, Italy, the Netherlands, Switzerland), North America (USA) and Asia (Japan, Malaysia) for more than 20 years. Different technical solutions, which vary primarily in the method of oxygen supply to the landfill, have been used. The selection of a relevant solution depends mainly on the final intention of its application, landfill construction, available financial resources, and legislation. In Europe, aeration is usually applied as a method for an acceleration of old landfill stabilization, commenced when the landfill gas production becomes insufficient for profitable energy recovery. Such an activity shortens the period of the residual gas production. This gas is usually released to the atmosphere because of problems with its combustion. The accelerated stabilization of waste, by a few-week aeration, is also used as the initial phase preceding of landfill mining. It allows to obtain an evident improvement in sanitary conditions before the workers will enter.

In the United States, the concept of aerobic bioreactor landfill where the wastes are subjected to the active aeration from the early phase of landfill lifetime, was developed. Additionally, in order to accelerate the waste stabilization, the leachate recirculation is used. In Asia, the other solution for the reduction of methane production inside the landfill was elaborated. It consists in landfill self-aeration caused by the differences in inner and outer landfill temperatures (Fukuoka method).

4.2 FUNDAMENTS OF THE AEROBIC DECOMPOSITION OF ORGANIC MATTER IN LANDFILL

Decomposition of organic matter under aerobic conditions results in the production of simple mineral compounds, i.e. carbon dioxide and water, but also humic-like substances (Ritzkowski et al. 2006). However, due to the heterogeneity of waste structure, the oxygen does not reach in the same amount all the places in landfill. Its concentration is determined by the waste moisture and porosity. As a result of oxygen diffusion impairment methanogenesis, denitrification, and sulphate reduction occur. Moreover, there are the processes such as the oxidation of an organic matter, the nitrification or the oxidation of methane that also take place there. Therefore, in the gas generated in aerated landfills CH_4, NH_3 and H_2S are present beside of the CO_2, CO or water vapor.

The decomposition of organic matter, under the presence of oxygen, starts from the hydrolysis and the oxidation of organic compounds (mainly sugars and proteins) by mesophilic microorganisms. As a result, the organic acids (causing a decrease in pH), CO_2 and H_2O

45

occurred. Aerobic decomposition processes are more exoergic than anaerobic ones (Read, 2001). Thus, an intensive degradation of the products of hydrolysis and oxidation cause a rise in the temperature inside the waste body, even to 80°C. Under such conditions microorganisms belonging to moderate thermophiles, with the optimum temperatures between 45°C–60°C, and strict thermophiles with the optimum temperature between 60–90°C (Kristjanss, 1991) begin to develop. Actinobacteria and fungi dominate among these microorganisms (Satyanarayana et al. 2013). They are able to digest moderately and hardly biodegradable compounds; such as lipids, hydrocarbons, cellulose, and hemicellulose. The main products of organic matter degradation in this phase are CO_2, H_2O, and NH_3. A depletion of substrates susceptible to biodegradation that occurred over the time results in a decrease in the temperature and the activity of mesophilic microorganisms appears again. Oxidation of NH_4^+ ions leads to a decrease in pH value of leachate. According to standards for composing, the oxygen concentration in air pores should not drop below 10% v/v (min. 5%) to maintain the efficient oxidation of organic matter. The rate of oxygen delivery must be maintained on an appropriate level in order to prevent the decrease of oxygen concentration due to its consumption by microorganisms. In the laboratory study carried out on waste taken from old landfill at the low aeration rate of about 0.0009 dm^3 kg^{-1}_{waste} min^{-1} the O_2-concentration lowered in the beginning phases below 5%, due to the intensive respiration of microorganisms; and grew to 15% during the further course of aeration (Prantl et al. 2006). However, this aeration rate would not be adequate for fresh organic-rich waste. The results of laboratory study on fresh solid waste carried out by Sekman et al. (2011) suggest that aeration rate of 0.10 dm^3 kg^{-1}_{waste} min^{-1} is sufficient and the most viable because of economic reasons. Based on the analysis of leachate (pH, chloride, alkalinity, COD, BOD$_5$, TKN, and NH_3-N) and waste settlement, they stated that no significant effect of the aeration rate in the range of 0.1 to 1.0 dm^3 kg^{-1}_{waste} min^{-1} was observed.

The optimum moisture content for the bacteria growth is within the range of 50 to 65% by weight. It was observed that the activity of microorganisms decreased by several orders of magnitude below a moisture content of around 40% (Read et al. 2001b). Therefore, it is recommended to maintain the moisture content in an aerobic landfill at least 40% and at most 70% to ensure the water availability for microbes and appropriate conditions for the oxygen diffusion. An excess of the upper limit value may deteriorate oxygen movement in waste pores and the process may tend to become anaerobic (Law et al. 2011).

Nakasaki et al. (1993) revealed that the optimum pH for a growth of microorganisms responsible for the aerobic degradation of proteins was within the range of 7–8, while decomposition of glucose proceeded rapidly in the pH ranged from 6 to 9.

4.3   CONSEQUENCES OF IN SITU LANDFILL AERATION

An oxygen presence accelerates the decomposition of organic matter in waste; and directs it to other pathways, which affects the quality of the products. The distribution of carbon load during the aerobic decomposition of organic matter is concentrated on the creation of gaseous products. Laboratory studies carried out in the aerated lysimeters filled with waste taken from an old landfill showed that after the 19-month experiment (when >97.5% of the maximal stoichiometric carbon discharge has been reached), 84% of the total carbon load was discharged in the form of CO_2. About 12% was discharged in a form of CH_4 and only 4% *via* the leachate (Ritzkowski & Stegmann, 2013).

4.3.1   *Landfill gas composition*

In situ aeration of waste leads to a significant change in the gas composition and production when compared to anaerobically produced LGF. In the laboratory study, the conversion of conditions, from anaerobic to aerobic, significantly influenced the quality and the quantity of landfill gas. After the start of aeration the methane production almost completely decreased within 1 month. More than 90% of carbon load transformed to gas phase was discharged as carbon dioxide. Meanwhile, carbon discharged as CO_2 in the anaerobic conditions was 40–50% of total carbon load. Additionally, the aeration enhanced carbon discharge. It was approximately 5-times higher compared to anaerobic conditions (Prantl et al. 2006).

According to the result of the field study conducted at a MSW landfill in Florida, air injection to the waste body caused a significant decrease in the ratio of CH_4 to CO_2. When the ratio of CH_4 to CO_2 decreased, the CO concentrations increased. However, air injection did not have any noticeable influence on VOCs and N_2O concentrations (Powell et al. 2006).

Due to the aeration, the ratio of CO_2 to CH_4 grows from *ca.* 0.5 that is typical for anaerobic conditions to 2.5–6 (Ritzkowski & Stegmann, 2007). This statement complies with the results of the field study carried out by Cossu et al. (2007). The authors noticed that after the start of the aeration on the old landfill, the methane concentration lowered from *ca.* 60% to *ca.* 2% and the ratio of CO_2 to CH_4 rose from 0.6 to 6. At least 80% decrease in CH_4 concentration in landfill gas was reported by Read et al. (2001a) on the pilot cell of Live Oak Landfill (USA) within 3 weeks of the waste aeration. The concentration of CH_4 remained below 15% v/v. A similar decrease in CH_4 concentration was observed at Milmersdorf landfill (Germany) after the onset of low pressure aeration. Before the start of the waste stabilisation, the concentrations of CH_4 and CO_2 in the LGF measured in the gas wells were in the range 50–80 vol.% and *ca.* 20 vol.%, respectively. After the aeration system was commenced, the CH_4 concentration rapidly decreased to 3–15%, and CO_2 was in the range of 10–20% (Heyer et al. 2005a).

The aeration causes the decrease in the concentration of H_2S, NH_3 and other trace gases (Powell et al. 2006). A drop of these gases content in LFG results in a reduction of odour nuisance associated with the waste disposal. It was reported that odours generated at landfills converted into aerobic objects were less pungent (Read et al. 2001b).

4.3.2   *Quality and quantity of landfill leachate*

4.3.2.1   *Changes in pH value*
The aeration causes a rapid increase in the pH value of the leachate, and keeps its value above pH 7. This is evidenced by the results of laboratory tests that were conducted in simulated landfill bioreactors (with leachate recirculation) filled with waste with varied initial pH. Erses et al. (2007) observed that the pH value in aerobic reactor increased from its initial value equal to 5.83 to the neutral values in only few days. Then, the pH values were reached between 7.5 and 8.0 until the end of the experiment. In anaerobic bioreactor, the pH values remained at the level about 5.5 within over 14 months, and then increased sharply to 7.0 after the onset of methanogenic activity. Significantly higher pH value in leachate from the aerobic bioreactor were observed also by Sang et al. (2008) during 140-day laboratory experiment. The pH value in aerobic reactor rapidly grew of *ca.* 2 pH units (from initial value equal to 7 to 9) within 3 weeks, and then it stabilized between the values 7.5–8. Meanwhile, in anaerobic reactor the pH values decreased to 6 within 3 weeks, and then it became gradually increasing, reaching the value of 7.5.

During 150-day laboratory experiment on solid waste stabilization in simulated landfill bioreactors, pH of leachate in aerobic reactor rapidly increased from 5–6 to *ca.* 7 during 40 days, and then it was gradually growing to values of 8–9, until the end of the study. The pH values of leachate of the anaerobic reactor was less than 5.5 until the day 100, which reflected the accumulation of volatile acids generated in an acidogenesis phase. Next, the pH value slightly increased to *ca.* 6 at the end of the experiment (Sekman et al. 2011).

4.3.2.2   *Reduction in leachate organic strength*
The results of laboratory and field studies show that the aeration of waste deposited in landfills causes an accelerated reduction of the organic matter contained in the leachate compared to the anaerobic reactor. Chemical oxygen demand (COD), total organic carbon

(TOC), biological oxygen demand (BOD) were used as the indicators of organic strength of leachate. Erses et al. (2008) showed about 96% reduction in COD, 85% reduction in TOC, and even 100% reduction in BOD in simulated aerobic landfill bioreactor after 12 months of the experiment. After this time in anaerobic reactor, COD and TOC were higher than initial values. The authors observed that in an initial phase of the experiment, the leachate produced under aerobic conditions was characterized by lower values of COD and TOC than the leachate from anaerobic reactor. The values of COD in leachates from aerobic and anaerobic reactors amounted to 17,900 mg dm^{-3} and 38,000 mg dm^{-3}, while the values of TOC were 1438 mg dm^{-3} and 17,990 mg dm^{-3}, respectively. The waste aeration resulted in the COD reduction to the level of 678 mg dm^{-3} after about 12 months; while in the anaerobic reactor, the COD value significantly increased. After a year, there was a sharp decline noted but it still remained at a high level. After 21 months, the COD in the anaerobic reactor was 900 mg dm^{-3}. A similar pattern was found in the case of TOC evolution. The value of this parameter in the aerobic reactor dropped to 218 mg dm^{-3} after approximately 12 months, while in the anaerobic reactor it reached 290 mg dm^{-3} after 21 months of the experiment.

Giannis et al. (2008) showed that the simulated aerobic landfill bioreactor could remove above 90% of chemical oxygen demand (COD) and almost to 100% of biochemical oxygen demand (BOD_5) from leachate. The conversion of anaerobic to aerobic landfill (with leachate recirculation) performed in field scale reduced biological oxygen demand (BOD) of leachate up to 70% within 3 months of aeration (Read et al. 2001b).

The BOD_5/COD ratio—the measure of biodegrability of organic matter—reached the value 0.03 after one-year of experiment in the simulated aerobic landfill reactor, and *ca.* 0.8 in anaerobic one. The BOD_5/COD ratio determined in the anaerobic reactor at the end of the experiment (after 630 days) was *ca.* 0.15 (Erses et al. 2008). Borglin et al. (2004) claimed that BOD_5/COD ratio in leachate from aerobic and anaerobic reactors that had worked 365 days were 0.03 and 0.45, respectively.

4.3.2.3   *Enhanced ammoniacal nitrogen removal*

During the microbial decomposition of waste, a part of organic nitrogen is converted to ammonium *via* ammonification processes. In anaerobic landfills, the reduction of ammonium in the leachate to environmentally acceptable levels takes from decades up to several centuries (Krümpelbeck, 2000; cited after Prantl et al. 2006). Oxygen presence enables the rapid ammonia removal via oxidation to nitrite and nitrate by nitrifying bacteria (*Nitrobacter* and *Nitrosomonas*). It causes even several-fold decrease in ammoniacal-nitrogen (NH_3-N) concentration in leachate from aerobic landfill in comparison with anaerobic one. Sekman et al. (2011) observed that after 150-day laboratory experiment NH_3-N concentration in leachate generated in the anaerobic reactor was around 1200 mg dm^{-3}, while in the aerobic reactor it was below 100 mg dm^{-3}. Mertoglu et al. (2006) reported that NH_3-N decreased from 250 to 30 mg dm^{-3} within 250 days of bioreactor aeration. What is more, Erses et al. (2008) also observed the rapid decrease in ammoniacal nitrogen concentration in aerobic laboratory bioreactor. The concentration of NH_3-N fell from *ca.* 400 mg dm^{-3} to 5 mg dm^{-3} even after 250 days, reaching the reduction level of approximately 99%. These findings were coherent with the results obtained by Cossu et al. (2003) in a laboratory experiment carried out on the waste from MBT plant. The NH_3-N concentration in the leachate declined from the initial value of about 1100 mg dm^{-3} to 5–6 mg dm^{-3} after 120 days of operation in the aerobic reactor. In contrast, NH_3-N concentration in a leachate from the anaerobic reactor increased to about 1000 mg dm^{-3} in a few weeks and remained at that level until the end of the experiment, which lasted for 630 days (Erses et al. 2008).

The time-dependent changes in total Kjeldahl nitrogen (TKN) in aerobic and anaerobic bioreactors run according to the similar pattern. The final reduction of TKN in aerobic reactor was *ca.* 93% (Erses et al. 2008), while in the anaerobic reactor the TKN value remained at the high level to the end of the experiment.

4.3.2.4   *Changes in leachate alkalinity*

The studies carried out in simulated landfill bioreactors by Sekman et al. (2011) showed that alkalinity levels in the aerobic and anaerobic reactors were >3500 mg $CaCO_3$ dm^{-3} from the beginning to the end of the experiment. Alkalinity in both types of reactors showed considerable fluctuations.

In earlier stages of waste decomposition, higher alkalinity level (even up to 9000 mg dm^{-3}) were found in leachate from the aerobic reactor. However, after about 80 days, the alkalinity in anaerobic reactor began to exceed the values measured in the leachate from the aerobic reactor, and at the end of experiment it was *ca.* 40% higher. Significantly lower alkalinity in the leachate from the aerobic reactor, when compared with the anaerobic one, was observed also by Erses et al. (2008) during a long-term study (630 days). In the aerobic reactor, alkalinity first increased to the highest value of 4500 mg dm^{-3} and then began to gradually decrease reaching even the value of 660 mg dm^{-3} at the end of experiment. In the anaerobic reactor, leachate alkalinity varied between 5500 and 7500 mg dm^3 and remained relatively constant during the acid formation phase. After the transition to methanogenic conditions, the pH values increased and total alkalinity decreased, reaching the value of 3800 mg dm^{-3} at end of the experiment.

4.3.2.5   *Decrease in heavy metals concentration*

The leachate recirculated through the aerobic bioreactor landfill affect the immobilization of heavy metals in waste. Such a conclusion can be derived from the results of laboratory and field study. According to Giannis et al. (2006), nickel and lead were the most efficiently removed among the 5 analysed heavy metals (Table 4.1.). After 8 months of the experiment, the concentrations of Ni and Pb decreased by 98% and 69%, respectively. A high reduction in lead and iron concentrations was observed in a field scale research. Lead was reduced to background levels and iron was reduced by 75–90% (Read et al. 2001b). The attenuation of heavy metals concentration in leachate recirculated through waste is related to high pH of the leachate, which favours the retention of heavy metals by various processes; such as sorption on solid phase, or carbonate and hydroxide precipitation (Giannis et al. 2008).

4.3.2.6   *Decrease in ecotoxicity of leachate*

Bio-essays indicated that leachate recirculated through the simulated aerobic bioreactor examined at the end of the experiment was not toxic for freshwater organisms. A toxicity test was performed on bioindicator *Daphnia magna*. In this test young Daphnia are exposed to various concentrations of the tested substance. The test involved the determination of the concentration of a substance in a water solution, where a half of the number of bioindicators dies. This value is called a lethal concentration (LC_{50}). It was stated that toxicity of the examined leachate dropped to zero after one year of the experiment (Table 4.2). That means that the chance to survival of the bioindicators in undiluted leachate was over 50% (Giannis et al. 2006).

Table 4.1   Concentration of selected heavy metals in leachate recirculated through a simulated aerobic landfill bioreactor (Giannis et al. 2006).

Metal	Unit	After 30 days	After 240 days	After 510 days
Ni	ppb	88.85	1.25	<1
Pb	ppb	54.97	17.05	<1
As	ppb	9.30	2.10	<1
Zn	ppm	9.46	1.25	<1
Cd	ppb	4.06	1.25	<1

Table 4.2   Results of the toxicity test with *Daphnia magna* conducted on leachate recirculated in simulated aerobic bioreactor landfill (Giannis et al. 2006).

Time of aeration [days]	Lethal concentration LC_{50} in 24 hours test [percent of leachate in diluted sample, %]
120	76.8
240	80.7
360	–

4.3.2.7   *Increase in chloride concentration*

Waste aeration causes the growth of chloride concentrations in leachate that were recirculated through the waste layer. It was observed that Cl⁻ concentrations in leachate began to increase simultaneously with the rise of the pH value. It can be explained by the influence of pH on chloride dissolution, which increases with the increase of pH (Biligili et al. 2006, Sekman et al. 2011).

4.3.2.8   *Reduction in leachate volume*

Landfill conversion from anaerobic to aerobic leads to the reduction of leachate volumes, greater than 80%, due to the intensive evaporation related to the increasing temperature inside the aerated landfill (Read et al. 2001b).

4.3.3   *Deposited waste parameters*

4.3.3.1   *Landfill settlement*

Aerobic bioreactor landfill technology enhances the rate and the extent of settlement. It was confirmed by both the laboratory and field studies. Erses et al. (2008) noticed that the settlement achieved in the simulated aerobic bioreactor was about 37% after 374 days, while in the anaerobic bioreactor it was only about 5% after 630 days. Giannis et al. (2008) reported 26% waste settlement in laboratory aerobic bioreactor (with leachate recirculation) after 510 days.

Waste settlements measured in the field scale were lower than the results obtained in the laboratory tests. After 2-year aeration of old Kuhstedt landfill (Germany) the settlement or subsidence was between 15 and 70 cm, which stands for 2% and 10% regarding the landfill height. Before the aeration was started, the considerable settlements of the landfill were found to be within the range of 10 cm (Heyer et al. 2005a). Waste settlement observed at two landfills in the USA after the aeration period of 9–18 months was 9–10% (Read et al. 2001a). The experiences from an *in situ* aerated old landfill showed that the major part of settlement appears during the first 18 months of system operation (Ritzkowski et al. 2004).

4.3.3.2   *Reduction of organic matter content in waste*

Aeration of waste causes an efficient degradation of organic matter that results in a higher reduction of the mass of organic substance. According to the laboratory experiment conducted by Erses et al. (2008) the reduction of organic substance content, achieved in the aerobic bioreactor, was 46%; while in the anaerobic bioreactor it was only 35%.

4.3.3.3   *Temperature inside the landfill*

During the aeration, the temperature inside the landfill increases due to the release of the energy from aerobic decomposition of organic matter. The result of field scale study showed that the average landfill temperature increased from about 30°C to 50°C within the first 12 months of aeration. This growth was in a good correlation with the waste settlement/subsidence. From this time, waste temperatures started decreasing. After 6 years, the temperature fell to the value that was measured before the trial. The decline continued also when the aeration was stopped (Ritzkowski & Stegmann, 2013). Heyer et al. (2005b) reported considerable rises in temperature up to the thermophilic range (50–70°C), which is common for composting. They were noted in a few aerated sections of the Kuhstedt and Milmersdorf landfills over several months. Furthermore, Read et al. (2001a) observed that temperature inside the aerated landfill were within the range of 40–60°C.

4.4　CONCEPTS OF LANDFILL AERATION

4.4.1　*Aerobic bioreactor landfill*

In an Aerobic Bioreactor Landfill (ABL) technology, the stabilization of waste is enhanced by the creation of aerobic conditions, and the optimum water conditions that are provided for bacteria growth. These goals are realized by air injection into the waste mass and liquids (mainly landfill leachate) recirculation through the layer of deposited waste (Fig. 4.1). Leachate is removed from the bottom layer, piped to liquids storage tanks, and recirculated into the landfill in a controlled manner (Bioreactor report). Landfill aeration may be realized on the newly designed objects (as-build landfill), that are destined for this type of waste deposition or on the conventional landfills, which had to be modified. In as-built bioreactor landfill the specific construction of the bioreactor components is implemented while waste is actively deposited. The second type of sites is a retrofit bioreactor landfill, in which all the necessary changes in the construction and the operation of bioreactor components occurs after the active operation phase, when waste was placed. For as-built bioreactor more choices for selecting the liquid addition techniques are available in comparison to retrofit bioreactors (Reinhart, 2012).

Pump and Treat Aerobic Flushing Bioreactor Landfill (PTAFBL), in which recalcitrant organics are oxidized both *in-situ* and *ex-situ* to the landfill (Fig. 4.2) is the new modified version of an aerobic landfill. This type of aerobic bioreactor was tested in pilot-scale by Bolyard et al. (2013). The Fenton's reagent was used for ex-situ leachate treatment because of its high efficiency in the removal of COD from leachate (Batarseh et al. 2007). The leachate containing oxidized organic compounds is returned to the landfill, where it was treated

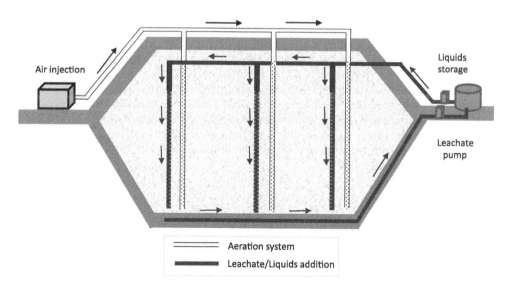

Figure 4.1　Concept of aerobic bioreactor landfill (EPA).

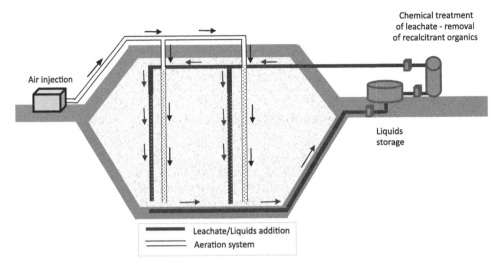

Figure 4.2   Concept of pump and treat aerobic flushing bioreactor landfill (Bolyard et al. 2013).

aerobically *in situ*. After this treatment, the ultimate end products remaining in the landfill were essentially humic matter, insoluble non-biodegradable organics, and immobilized inorganic compounds (Bolyard et al. 2013).

4.4.2   *Semi-aerobic landfill*

In this solution, an aeration is a consequence of the temperature differences between waste and the atmospheric air. Due to the exoergic nature of the organic matter decomposition, the temperature inside the landfill is higher than the ambient air. Therefore, the warm and low-density landfill gas migrates upwards the waste body via the venting pipes. This warm gas is replaced with a colder ambient air, which is sucked through leachate drainage pipes, placed in the bottom part of the landfill (Fig. 4.3). The pipes are designed in a way that only one-third of the section is filled with liquids (Ritzkowski & Stegmann, 2012). Leachate drainage system fulfils a dual role. It is also used as an air supply system. Such a solution of the landfill aeration is practiced in some Asian countries. The first pilot semi-aerobic landfill was constructed in Fukuoka City (Japan) in 1975. After ascertaining the benefits of this method of the waste disposal, it was officially recommended in the Final Waste Disposal Guidance issued by the Ministry of Health and Welfare of Japan in 1979, as a Fukuoka method (Tashiro, 2011).

The main aim of releasing this technology was to promote the leachate aerobic stabilisation. The fundamental advantage of this method is a low operating cost. It does not require expensive air pumping. However, it does not provide the waste body with an adequate delivery of oxygen, which results in only a partial reduction of methanogenesis. Methane concentration in the off-gas from semi-aerobic landfill decreases to 20–30% (Tashiro, 2011; Ritzkowski & Stegmann, 2012).

4.4.3   *In-situ aeration of old landfills*

In recent years, there has been an increased interest in the stabilisation of old landfills. On the one hand, these landfills do not generate enough methane to ensure the profitable LFG recovery. On the other hand, they still pose a threat to the environment and must be monitored. The acceleration of the decomposition of the hardly biodegradable compounds remaining in waste allows to shorten the post closure care period. This method is also used as a preparation phase before the waste mining or landfill recultivation.

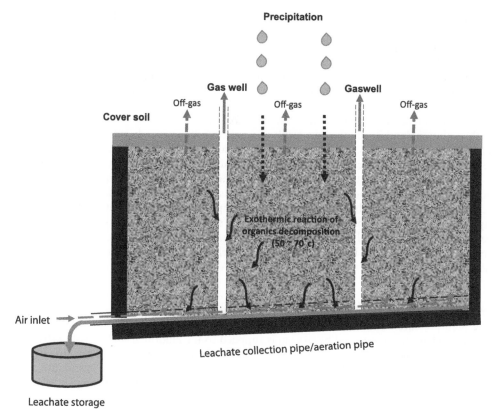

Figure 4.3   Concept of semi-aerobic landfill (Tashiro, 2011).

4.5   METHODS OF AIR SUPPLY TO THE LANDFILL

The main criterion to distinguish different technological solutions of landfill aeration is the way in which the air is supplied to the landfill. The air fed into the waste layer may be forced by the change in pressure (or sucking air injection), or temperature. Detailed overview of existing solutions for landfill aeration, their application ranges and specifications are given by Ritzkowski & Stegmann (2012). When it comes to the present paper, the main principles of the currently known solutions in the scope of landfill aeration are described.

4.5.1   *Low pressure aeration*

In these solutions, the air is introduced into a landfill on the basis of pressure differences, which do not exceed 0.3 bars. They usually are within the range of 20–80 mbars (Ritzkowski & Stegmann, 2012).

4.5.1.1   *Active aeration without or with off-gas extraction*
In active aeration systems, the compressed atmospheric air is continuously injected to the landfill body, usually through a system of gas wells. The air migrates in the porous waste medium by means of convection and diffusion. In the system working without the off-gas extraction, the low-contaminated gas is collected and treated in a controlled manner by means of other wells (Heyer et al. 2005b) or in other cases gas can migrate through the landfill cover, which acts as a biological filter for landfill gas treatment (Ritzkowski & Stegmann, 2012).

In the concepts operating with the off-gas extraction, the gas is sucked out by simultaneously operated off-gas extraction system, which consists of gas wells connected to an air compressing unit (Ritzkowski & Stegmann, 2012). Then, the off-gases can be purified with the use of biological or thermal methods. The active aeration systems were applied inter alia at the landfills in: Kuhstedt, Milmersdorf, Schwalbach-Griesborn (Germany), Mannersdorf, Pill, Heferlbach (Austria), Sassari, Legnago (Italy) (Ritzkowski, 2011).

4.5.1.2   *Passive aeration (air venting or over-suction system)*

In this solution, landfill aeration is a result of a negative pressure induced inside the landfill. The atmospheric air is sucked into waste over the surface of the landfill and/or *via* passive aeration wells. Gas extraction rates significantly exceed the LFG formation rates, leading to a gradual aeration starting from the landfill surface towards the deeper areas. The aeration process is relatively slow and it is difficult to ensure that the entire waste mass becomes homogenously aerated. Therefore, this method is implementable only at sites with the waste layer thicknesses of <10 m as, wherein, the oxygen supply; and thus, aerobisation may not be guaranteed (Hupe et al. 2003). The exemplary landfills, at which air venting system was applied, are the following: Kiel Drachensee, Schenefeld (Germany), Sass Grans (Switzerland) (Ritzkowski, 2011).

4.5.2   *High pressure aeration*

The high pressure BIOPUSTER® method was developed by PORR Umwelttechnik. The compressed air or air enriched by oxygen is supplied to the landfill at high pressure (from 3 to 7 bars) from lances. Air is not injected to waste with constant rate but in pressure waves pulsation. High pressure and high oxygen concentration (event at 40 vol.%) ensure the oxygen penetration both in highly and weakly compacted waste materials, and completely aero-

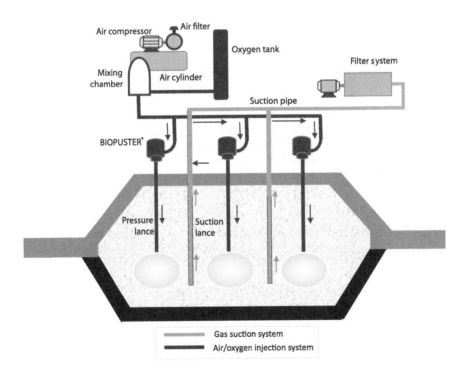

Figure 4.4   Gas venting system in BIOPUSTER® (MBU website).

bic conditions. The off-gas is extracted by suction lances and more extracted air is sucked out than fresh air was injected (Spillmann, 2001). The extraction capacity is of 30% higher than gas volumes used for aeration. The off-gases are usually treated by biofiltration (Biopuster, MBU). The concept may be used both for the fresh and old waste stabilization but it is mainly associated with the implementation of landfill mining projects (Ritzkowski & Stegmann, 2012). The scheme of BIOPUSTER® system is presented in the Figure 4.4. The system was successfully implemented, for example, in the old municipal area of Feldbach (Styria, Austria) in the seventies of the 20th century. It was the first *in situ* landfill restoration where a controlled aerobic waste stabilization has ever been used. Furthermore, the injection of warm water enriched with microorganisms was applied through the pressure and additional lances. Other examples of BIOPUSTER®-System application is "Helene Berger" landfill. After 12 weeks of the aeration, the transition of conditions from anaerobic to aerobic was completed, and the excavation of waste was done. However, only a quarter of the excavation materials were re-deposited (Budde et al. 2002).

4.6 ADVANTAGES AND DISADVANTAGES OF THE LANDFILL AERATION

Regardless of the applied technological solution, the decomposition of an organic matter in the presence of oxygen leads to many benefits when compared to the anaerobic conditions. The most essential advantages are:

- reduction of the negative impact of landfills on the atmosphere due to decrease in methane concentration with GWP 25 times higher than that of carbon dioxide. According to a model calculation done by Ritzkowski & Stegmann (2007), more than 72% of the total GHG emissions occurring under anaerobic conditions in the landfill could be avoided by altering the aerobic conditions;
- reduction of the negative impact of landfills on the underground water and soil mainly due to changes in the quality and quantity of leachate;
- shortening of the landfill stabilization period resulting from the acceleration of organic matter decomposition;
- enhancement of waste settlement resulting from more rapid and more efficient decomposition of organic matter;
- elimination of explosion risk due to the reduction of gas pressure (e.g. in the case of semi-aerobic landfill) and the significant decrease of methane concentration;
- attenuation of odour nuisance resulting from the decrease of H_2S, NH_3 and other malodorous compounds;
- reduction of land consumption for waste disposal due to the better use of the volume of existing landfills;
- reduction of final cost of waste landfilling resulting from the better use of landfill volume (possible recovery even to 30% of volume), shortening of post closure care period, elimination or limitation of leachate treatment problem, and the limitation of the risk of landfill gas escape (lower cost of landfill capping system).

4.7 CRITICAL APPROACH TO THE LANDFILL AERATION CONCEPT

Although the landfill aeration offers many benefits for the environment, there are also some unresolved problems, which should be mentioned. What is more, not all of the consequences of waste aeration have been thoroughly understood.

Ritzkowski (2011) pays attention to uncertainty regarding the actual amount of N_2O emissions. Aerobic conditions favour the generation of this gas, which has a high GWP about 298 (in 100 years). Although, the increase of N_2O production was not observed in field scale study during the landfill aeration (Powell et al. 2006), the lab-scale experiment showed much higher emission of N_2O from aerated bioreactor when compared with the anaerobic one. This effect was the most evident when the partially stabilised waste was digested. Therefore, the intensification of N_2O emissions from old landfills that are subjected to aeration could be expected.

An intensive discharged of organic carbon load during aerobic biodegradation of organic matter, accumulated in the waste, is another problematic issue. On the one hand, it can be considered as a positive effect. On the other hand, the role of landfill as a final repository of carbon is limited in this way. The compounds, such as lignin and hemicelluloses, which are hardly or even non-biodegradable under anaerobic conditions became susceptible to a microbial decomposition in oxygen-enriched atmosphere.

Another issue that is worth mentioning is the lack of generally accepted criteria on which the assessment of waste stabilization success could be based. Although, a number of pilot projects related to landfill aeration have already been implemented, such a rules have not yet been developed. Ritzkowski & Stegmann (2013) have proposed 6 criteria to determine the measures of landfill aeration completion (Table 4.3).

An operational problem that occurs in low pressure aeration solutions regards the difficulties in maintenance of the aerobic conditions inside the entire waste mass. The anaerobic areas were observed even at the low moisture content of waste equal to 33% (Yazdani et al. 2010), which should definitely favour the gas migration in porous medium.

From practical point of view, the main disadvantage of landfill aeration is an increased cost of its application, especially in the case of small landfills that had to be retrofit. According to the review done by Ritzkowski & Stegmann (2012), it is estimated that the cost of low pressure aeration system with parallel aeration and off-gas extraction is between 0.45 and up to 7 € per 1 m^3 of landfilled waste, depending on various factors, such as existing infrastructure and landfill volume. One of the elements of aerobic landfill equipment that generate cost is the station for the off-gas treatment. This gas cannot be used for an energy purpose and directly flaring because of the low CH_4 concentration. Nevertheless, such off-gas still contains some amount of methane and other compounds, which are not neutral for the environment. Therefore, its treatment is required. Furthermore, the air injection/suction causes the increase in the volume of the off-gas that should be purified.

The contaminated off-gas may be treated by means of thermal (adsorption on activated carbon, non-catalytic, and autothermic methods) or biological methods (bio-washers and biofilters) (Heyer et al. 2005a).

Table 4.3   Criteria and target values to determine the completion of landfill aeration measures (Ritzkowski & Stegmann, 2013).

Criteria	Value or dimension	Description
Conversion rate of biodegradable organic carbon (BOC)	C discharge, aerated LF ≥ 90% BOC (Mg C)	Determined by permanent off-gas analysis (CO_2 and CH_4)
LFG generation	CH_4 formation ≤ 0.5 dm^3/m^2h	On the base of LFG extraction test conducted after a period of 3 weeks without active aeration
LFG generation potential (GP_{21})	GP_{21} ≤ 10 dm^3/kg DM	Laboratory test on waste taken from the aerated landfill
Respiratory index (RI_4)	RI_4 ≤ 2.5 mg O_2/g DM	Laboratory test on waste taken from the aerated landfill
Landfill settlements	Significant reduction	
Landfill temperature	Average temperature in a range comparable to the situation before aeration	Online measurements inside landfill at different depths and areas

It should be borne in mind that considerable part of the cost incurred for the operation of aerated landfill may be compensated by cost-saving aspects, which were mentioned in the previous paragraph.

4.8   SUMMARY

According the current state of knowledge, landfills aeration leading to in-situ composting of deposited waste is a method that allows for the effective elimination of a long-term impact of landfills on the environment. It contributes to the attenuation of the role of landfill in a formation of greenhouse gas effect by preventing the production of methane; which is the basic component of the gas release from a conventional landfill. However, the influence of waste aeration on N_2O production poses the important question within the scope of aerated landfill role in GHG emissions. This have to be clarified in the very near future.

The universality of the aeration applicability, which results from the wide selection of technical and technological solutions, is a significant advantage of this method. It can be implemented into the existing landfill sites in various stages of its operation, where it may serve as a method to accelerate the stabilization of waste. It can be also applied as the target solution to the new-built objects. However, the transformation of anaerobic landfills into aerobic sites requires the implementation of a cost-intensive modifications; involving the construction of the systems to supply, distribution, and usually the gas extraction as well as its treatment. Moreover, a long-term delivery of air to the waste body involves high operational costs. These are important factors limiting the practical use of the landfill aeration as a method of the waste deposition. However, taking practical premises, such as a limiting area for building new landfills or potential profits from leachate management, into consideration, the interest in the enhancement of landfill stabilization process can be expected.

REFERENCES

Batarseh, E., Reinhart, D. & Daly, L. (2007). "Liquid Sodium Ferrate and Fenton's Reagent for Treatment of Mature Landfill Leachate." *J. Environ. Eng.*, 133(11), 1042–1050.

Bilgili, M.S. Demir, A. & Özkaya, B. 2007. Influence of leachate recirculation on aerobic and anaerobic decomposition of solid wastes, *J. Hazard Mater*. 143(1–2): 177–183.

Bioreactor report (http://www.bioreactor.org/BioreactorFinalReport/FinalReport VOLUME1_10/ AttachmentforVOLUME8/Bioreactor_Landfill_OperationV10.pdf

Bolyard, S.C., Reinhart, D.R. & Santra, S. 2013. Pump and Treat Aerobic Flushing Bioreactor Landfill. Sinks a Vital Element of Modern Waste Management 2nd International Conference on Final Sinks 16–18 May 2013 Espoo, Finland.

Borglin, S.E., Hazen, T.C., Oldenburg, C.M. & Zawislanski, P.T. 2004, Comparison of aerobic and anaerobic biotreatment of municipal solid waste, *J. Air Waste Manag Assoc.* 54(7), 815–22.

Budde, F., Chlan, P. & Dörrie, T. (2002): Landfill restoration with the BIOPUSTER-System–Aeration as prerequisite for occupational-, residential and environmental safety. EUROARAB 2002, 10th–12th October 2002, Institute f. LBAW, University of Rostock.

Cossu, R., Raga, R. & Rossetti, D. 2003. The PAF model: an integrated approach for landfill sustainability. *Waste Management* 23, 37–44.

Cossu, R., Raga, R., Rossetti, D. & Cestaro, S., 2007. Case study of application of the in situ aeration on an old landfill: Results and perspectives, Sardinia 2007.

Debra, R. Reinhart, Pump and Treat Aerobic Flushing Bioreactor Landfill (Year 1), December 2012 (http://www.cee.ucf.edu/people/reinhart/research/PTAFBL/Documents/Pump%20and%20Treat%20 Aerobic%20Flushing%20Bioreactor%20Landfill%20Year%201%20Report%20Final.pdf).

Erses, A.S., Onayn, T.T. & Yenigun, O. 2008. Comparison of aerobic and anaerobic degradation of municipal solid waste in bioreactor landfills, *Bioresource Technology* 99 (13): 5418–5426.

Giannis, A., Makripodis, G., Simantiraki, F., Somara, M. & Gidarakos, E., 2008, Monitoring operational and leachate characteristics of an aerobic simulated landfill bioreactor. *Waste Management*, 28 (8), 1346–1354.

He, P., Yang, N., Gu, H., Zhang, H. & Shao, L.J. 2011. N_2O and NH_3 emissions from a bioreactor landfill operated under limited aerobic degradation conditions. *Environ Sci* (*China*) 23(6): 1011–1019.

Heyer, K.-U., Hupe K., Koop A. & Stegmann, R. 2005b. Aerobic in situ stabilisation of landfills in the closure and aftercare period. Proceedings Sardinia 2005, Tenth International Waste Management and Landfill Symposium S. Margherita di Pula, Cagliari, Italy; 3–7 October 2005 by CISA, Environmental Sanitary Engineering Centre, Italy.

Heyer, K.-U., Hupe, K., Ritzkowski, M. & Stegmann, R. 2005a. Pollutant release and pollutant reduction—Impact of the aeration of landfills, *Waste Management* 25 (2005) 353–359.

Kristjanss J.K. (Ed.). 1991. Thermophilic Bacteria. CRC Press.

Krümpelbeck, I. 2000. Untersuchungen zum langfristigen Verhalten von Siedlungsabfällen. Dissertation an der Gesamthochschule Wuppertal.

Law, J., Peterson, E. & Hudgins, M. Water requirement estimates for an aerobic bioreactor landfill in China. Proceedings Sardinia 2011, Thirteenth International Waste Management and Landfill Symposium S. Margherita di Pula, Cagliari, Italy; 3–7 October, 2011. CISA, Environmental Sanitary Engineering Centre, Italy.

MBU website (http://www.mbu.at/mbu/index.php/en/landfill-technology).

Mertoglu, B., Calli, B., İnanc, B. & Ozturk, I. 2006. Evaluation of in situ ammonia removal in an aerated landfill bioreactor. *Process Biochemistry* 41 (12), 2359–2366.

Nakasaki, K., Yaguchi H., Sasaki Y. & Kubota H. Effects of pH Control On Composting of Garbage *Waste Manag Res* March 1993 vol. 11 no. 2, 117–125.

Powell, J., Jain P., Kim H., Townsend T. & Reinhart D. 2006. Changes in Landfill Gas Quality as a Result of Controlled Air Injection. *Environ. Sci. Technol.*, 40(3): 1029–1034.

Prantl, R., Tesar M., Huber-Humer M. & Lechner P. 2006, Changes in carbon and nitrogen pool during in-situ aeration of old landfills under varying conditions, *Waste Management*, 26(4), 373–380.

Read, A.D., Hudgins M. & Philips P., 2001a, Perpetual landfilling through aeration of the waste mass; lessons from test cells in Georgia (USA), Waste Management, 21(7), 617–629.

Read, A.D., Hudgins, M., Harper, S., Phillips, P. & Morris, J., 2001b. The successful demonstration of aerobic landfilling: the potential for a more sustainable solid waste management approach? *Resources, Conservation and Recycling* 32, 115–146.

Ritzkowski, M., 2011. New project types in CDM waste sector: Landfill aeration. Current and future applications. Practitioners Workshop on CDM Standards 8–10 June 2011. Bonn.

Ritzkowski, M. & Stegmann, R. 2007. Controlling greenhouse gas emission by in situ aeration. International Journal of Greenhouse Gas Control 1(3): 281–288.

Ritzkowski, M. & Stegmann R. 2012. Landfill aeration worldwide: Concepts, indications and findings, *Waste Management* 32(7): 1411–1419.

Ritzkowski, M. & Stegmann, R. 2013. Landfill aeration within the scope of post-closure care and its completion. *Waste Management* http: //dx.doi.org/10.1016/j.wasman.2013.02.004

Ritzkowski, M., Heyer, K.-U. & Stegmann R., 2006. Fundamental processes and implications during in situ aeration of old landfills, *Waste Management* 26: 356–372.

Sang, N.N., Soda S., Sei K. & Ike M., 2008, Effect of aeration on stabilization of organic solid waste and microbial population dynamics in lab-scale landfill bioreactors, *J. Bioscience and Bioengineering* 106(5): 425–432.

Satyanarayana, T., Littlechild, J. & Yutaka Kawarabayasi, Y. (Eds.), 2013 Thermophilic Microbes in Environmental and Industrial Biotechnology biotechnology of thermophiles. Dordrecht; New York: Springer.

Sekman, E., Top, S., Varank G. & Bilgili, M.S. 2011. Pilot-scale investigation of aeration rate effect on leachate characteristics in landfills. *Fresenius Environmental Bulletin* 20(7a): 1841–1852.

Slezak, R., Krzystek L. & Ledakowicz S. Simulation of aerobic landfill in laboratory scale lysimeters—effect of aeration rate. *Chemical Papers* 64(2): 223–229.

Spillmann, P. 2001. Re-use of areas infilled with municipals waste and industrail residues by specific treatment. In: The exploitation of natural resources and the consequences. The proceedings of GREEN 3. The 3rd International Symposium on Geotechnics Related to the European Environment. R.W. Sarsby, T. Meggyes (eds). Berlin, Germany, June 2000: Thomas Telford. London: 2001196–202.

Tashiro, T. 2011. The "Fukuoka Method": Semi-Aerobic Landfill Technology. IRBC Conference, MetroVancouver 20–22 Sept. 2011. (http://www.metrovancouver.org/2011IRBC/Program/IRBC-Docs/IRBC-Factsheet_FukuokaMethodWasteMgmt_Fukuoka.pdf).

Yazdani, R., Mostafid, E., Han, B., Imhoff, P., Chiu, P., Kayhanian, M. & Tchobanoglous, G., 2010. Performance of enhanced aerobic landfill bioreactor under air flow and liquid addition, Abstracts of Papers, 239th ACS National Meeting, San Francisco, CA, United States, March 21–25.

CHAPTER 5

Biological oxidation as a method for mitigation of LFG emission

5.1 INTRODUCTION

The ability of microorganisms to use different compounds of an organic and inorganic nature, as a source of building elements and energy, reduces greenhouse gas emissions from landfill sites. Microorganisms; especially bacteria growing in the porous material, which constitute landfill cover, and also the one living in the surface layer; decompose some of the compounds contained in landfill gas. As a result of this decomposition, called biodegradation, simple mineral and organic compounds, which are less harmful to the environment than parent compounds, are formed. The initiation of this process does not require special inoculation of the material. Biodegradation of LFG components occurs spontaneously because these microorganisms are widespread in the environment. However, the creation of optimal conditions for their growth, by skilful selection off filling material, or the amount of incoming substrate increases the use of their abilities. It is also possible to introduce specialized microorganisms into the material, in which biodegradation takes place, which leads to the increase of the process rate or enhanced the efficiency of contamination removal.

Research on the use of microorganisms to reduce pollutant emissions from landfills were initiated in the 90s of the 20th century. Starting from that date an intensive development of investigations on use of biological method for landfill gas emission mitigation has been observed. The first who have become interested in this field were Whalen et al. (1990), Jones & Nedwell (1993), Bogner & Spokas (1993), Kightley et al. (1995), Czepiel et al. (1996), Kjeldsen et al. (1997). Increasing awareness of the need to protect the environment and technological development in environmental engineering have broadened the scope of research in this field. Initially, the research was carried out for the application of different types of materials for the oxidation of LFG components in landfill soil biocovers. Currently, since the isolation of waste from the environment is considered to be the main condition for "sustainable landfill," the role of landfill biocovers as filters of landfill gas treatment has been limited. This leads to the search of different kinds of solutions, which would be adapted to the technological and technical conditions at modern landfills. For instance, biofilters and biowindows may be the solutions. Although their action is based on the same process that takes place in landfill biocovers, a different technical preparation is still required. Biotarps constitute a new form of the use of microorganisms which aim to reduce the impact of landfills on the environment. Those are flexible mats, inoculated with microorganisms, and placed on a layer of waste for a short time, e.g. such as daily covers. When it comes to the data concerning field performance of biotarps, they are not available yet.

Methanotrophic microorganisms play a very important role in landfills. Microbial oxidation of methane in the upper layers of waste and landfill biocovers leads to a reduction of methane emissions from landfills by at least 25–30% per year (Nozhevnikova et al. 1993; Chanton & Liptay, 2000). The analysis of data on 42 landfills shows that the removal efficiency of methane in landfill biocovers is from 22% in clayey materials to 55%, in sandy materials. The overall mean value across all the studies was 36% ± 6%, while the range of annual fluctuations observed in the 15 landfills was from 11% to 89% (Chanton et al. 2009). In practice, the share of methanotrophs in the reduction of methane emissions from landfills is much larger than 10% which is assumed in the models of methane emissions.

The main advantages of biotechnological methods for mitigation of landfill gas emission are low operating and capital costs, low pressure drop, and no further waste streams. The disadvantages are low efficiency of removal of components with high concentration and sensitivity to atmospheric conditions.

5.2   FUNDAMENTALS OF MICROBIAL REMOVAL OF LFG COMPONENTS

Biotechnological methods of reducing the emissions of pollutants from landfills are based on the biochemical decomposition of biodegradable compounds in landfill gas while passing through the porous material inhabited by microorganisms. The process takes place under aerobic conditions. The microorganisms form a complex community (biofilm) attached to the surface of soil particles or are suspended in the water phase surrounding the particles (Fig. 5.1). The individual components of the gas are absorbed from the gas mixture to an aqueous phase, and they move by diffusion. The components may migrate in aqueous phase to the solid phase, where they can also be adsorbed. However, if they are in the immediate vicinity of microbial cells, they can penetrate the cell membranes to the interior, where they are metabolised. Through oxidative processes, contaminations are converted to carbon dioxide, water vapour, organic biomass, mineral salts, and other compounds. The microorganisms use some components of the landfill gas as a source of energy or carbon, which is essential for the growth of the population. It is facilitated due to having appropriate enzymes, which break the bond structure of the chemicals.

Decomposition of compounds by microorganisms results in the formation of simple mineral matters, which is a process of mineralization. If mineralization relates to compounds harmful to the environment, it is being called biodegradation. Rate and efficiency of biodegradation dependent on many factors including the number of the microorganisms, degree of their acclimation, accessibility of water, oxygen and nutrients, external and internal (cellular) transport properties, temperature, properties of the oxidized compounds, etc.

The potential vulnerability of a compound to biodegradation can be considered at two levels:

1. physical, determining the ability of the compound to transfer into the liquid phase; which is a prerequisite for the migration of the compound to the cells of microorganisms. The physical level is connected with the following properties of the compound, namely hydrophobicity, solubility in water and environmental parameters such as temperature, salinity or pH;
2. biochemical, determining the ability to metabolize a given compound by the microorganisms. This entails the presence of specific enzymes responsible for the disruption of different types of chemical bonds in certain chemical compounds and toxic nature of the degraded compounds relative to the microorganisms.

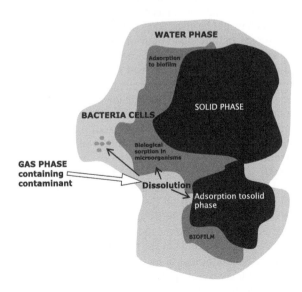

Figure 5.1   Fate of gas in biofilter medium (Devinny et al. 1999, modified).

Therefore, the susceptibility of a compound to biodegradation (i.e. biodegradability) depends on the structure of the compound, its physical and chemical properties; as well as the qualitative and quantitative composition of the microbial consortia.

Generally, biodegrability of the compounds decrease with a drop of water solubility and with an increase of molecular weight and complexity of bond structure (Devinny et al. 1999). The microbial removal of the compounds is limited by its hydrophobicity (lipophilicity), determining the water solubility and thereby pollutant transfer from the gas to the aqueous phase. The hydrophobicity is characterized by the octanol-water partition coefficients. It is a ratio of concentrations of the un-ionized compound in a mixture of two immiscible phases (octanol and water) at equilibrium. It determines the difference in solubility of the compound in these two phases. Estimation of hydrophobicity of the compounds is assessed on the base of logarithm of the ratio (logP). It is assumed, that the compound is hydrophilic

Table 5.1 Some properties of selected chemicals detected in landfill gas and their biodegrability (EPA ChSP, GESTIS substance database, ICSC database).

Compound	Molar mass [g mol^{-1}]	Solubility in water at given temperature [mg L^{-1}]	logP	Biodegrability[1] Weak	Mode-rate	Good	Un-known
Aliphatic hydrocarbons							
Methane	16.04	22.04 [20°C]	1.09	■			
Propane	44.10	70 [20°C]	2.236				■
Aromatic hydrocarbons							
Benzene	77.11	1800 [25°C]	2.13		■		
Toluene	92.14	470 [20°C]	2.73			■	
Ethyl benzene	106.17	140 [20°C]	3.15			■	
m-Xylene	106.16	174 [25°C]	3.20			■	
p-Xylene	106.16	180 [20°C]	3.15			■	
o-Xylene	106.16	174 [25°C]	3.12		■		
Styrene	104.15	310 [25°C]	3.0		■		
Chlorinated hydrocarbons[2]							
Carbon tetrachloride	153.82	793 [25°C]	2.64	■			
Chloroform	119.38	8 × 10^3 [20°C]	1.97	■			
Dichloromethane	84.93	13.2 × 10^3 [25°C]	1.25			■	
Vinyl chloride	62.6	2.76 × 10^3 [25°C]	1.58	■			
1,2-Dichlorobenzene	147.01	156 [25°C]	3.38				■
1,1,1-Trichloroethane	133.4	1.33 × 10^3 [25°C]	2.49			■	
Organic sulphur compounds[2]							
Carbon disulphide	76.139	1.19 × 10^3 [25°C]	1.84		■		
Dimethyl sulphide	62.13	22 × 10^3 [25°C]	0.977		■		
Dimethyl disulphide	94.2	2.5 × 10^3 [20°C]	1.77	■			
Methyl mercaptan	48.11	23 × 10^3 [20°C]	0.72	■			
Other compounds							
Acetone	58.08	1 × 10^6 [25°C]	−0.042			■	
Methyl ethyl ketone	72.11	275 × 10^3	0.29			■	
Hydrogen sulphide	34.08	4 × 10^3 [20°C]	−1.38			■	
Ammonia	17.031	310 × 10^3 [25°C]	n.d			■	

[1] According to Devinny et al. (1999).
[2] Treatment of these compounds may cause a change of filter bed pH, which may decrease an efficiency of these gases removal.
n.d—no data.

when logP < 0; moderate lipophilic when 1 < logP < 3, and highly lipophilic when logP > 3 (Takacs-Novak, 2012). Highly liphofilic compounds pose a serious threat to the environment because of the possibility to bioaccumulation.

Physical properties of selected components of landfill gas significant from the point of view of their susceptibility to microbial decomposition and evaluation of their biodegradability cited by Devinny et al. (1999) are presented in Table 5.1.

The data presented in Table 5.1 depict that low molar mass and solubility in water are not the only criteria for assessing the vulnerability of compound to biodegradation. Some of the compounds with high lipophilicity (logP > 3); for example ethyl benzene, xylenes are believed to be biodegrability good; whereas, benzene that is more soluble in water is poorly biodegradable. Liu et al. (2003) stated that biodegrability of 51 substituted benzenes was mainly affected by some substructures such as $-CH_3$, $-CH=$, $-NH_2$, $-NO_2$, $-OH$, $-SO_3H$, and Cl. The results of model study of Boethling (1986) suggest a possible role of molecular shape as a determinant of biodegradability. In contrast, studies conducted by Han et al. (2008) show a significant effect of cyclization to biodegradation. The authors claimed that the rate of biodegradation of different naphthenic acids significantly lowered with the increase of cyclization, but the carbon number in the particles had little effect on its biodegradation.

5.3   BIOOXIDATION OF METHANE UNDER AEROBIC CONDITIONS

In regard to qualitative and quantitative aspects, micro-organisms present in the landfill cover soils or biofilters, through which landfill gas passes, constitute a very diverse group. This is mainly due to the presence in the biogas numerous chemical compounds, which can be used by the different species of bacteria or fungi as a source of carbon and energy. A group of microorganisms, most frequently reported in the literature, that contribute to the mitigation of landfill gas are methane oxidizing bacteria (methanotrophs, methane-oxidising bacteria). Less attention has been paid to other microorganisms growing in landfill covers or biofilters, which are responsible for the removal of trace gases from landfill gas.

5.3.1   *Methane-oxidising microorganisms: Classification and habitat requirements*

Methane oxidizing microorganisms, called as methanotrophs are the prokaryotic microorganisms, whose have an ability to use methane, the simplest hydrocarbon, as a sole carbon and energy source (Fig. 5.2). They are a special subgroup of methylotrophs, bacteria that can use one-carbon organic compounds or multi-carbon compounds but containing no carbon bonds, such as dimethyl ether and dimethyl amine as the carbon or energy source. In the literature, information about the existence of eukaryotes—yeast, belonging to the kingdom *Fungi* that are able to metabolize methane; may also be found (Wolf & Hanson, 1979, Wolf & Hanson, 1980, Wolf et al. 1980). Researchers from Department of Bacteriology, University of Wisconsin stated that at least four different strains of yeast isolated from lake water and soil were capable of utilizing CH_4 as the sole energy source for growth. They can oxidize methane to CO_2, but the intermediates such as methanol, formaldehyde and formate would not support cell growth (Wolf & Hanson, 1979). But nothing is known of the enzymology of methane oxidation in these yeasts (Higgins, 1981). For instance, *Sporobolomyces roseus* strain Y and *Rhodotorula glutinis* strain CY, growing very slowly on methane, with a generation time of at least two days, belong to those fungi (Wolf et al. 1980). However, the existence of eukaryotic organisms capable of utilizing CH_4 as the sole carbon source has not been confirmed. Although a lot of them are classified as methylotrophs. Eukaryotic methylotrophs are limited to a number of yeast genera including *Candida*, *Pichia*, and some genera that were recently separated from *Pichia*; that is, *Ogataea*, *Kuraishia*, and *Komagataella*. They can only use methanol as a carbon source, and methylamine not as a carbon source but as a nitrogen source (Limtong et al. 2008).

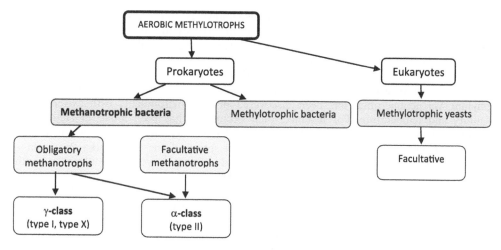

Figure 5.2   Taxonomic affiliation of methanotrophs (Trotsenko & Stępniewska, 2012).

In contrary, prokaryotic methanotrophs constitute a very numerous group of microorganisms. It consists of bacteria having the ability to grow on methane used as a source of carbon and energy. These microorganisms are involved in the carbon cycle. They convert it from the reduced form, in which it appears in oxidized organic compounds, into the oxidized form of CO_2. At the same time they play an important role in the environment by reducing the concentration of CH_4 in the atmosphere. They fulfil this task primarily by preventing the emissions of methane from the Earth's surface, oxidizing it during its migration from the source to the atmosphere. According to the estimations made by Reeburgh et al. (1993), it turned out that the biological activity of methanotrophs in a variety of natural and anthropogenic ecosystems reduces the methane emissions of about 700 Tg per year. This value is higher than the total annual methane emission from all sources amounting to about 582 Tg (IPCC, 2007). Methanotrophs are also capable of the oxidation of methane present in the atmosphere. In accordance with the data from the IPCC (2007) the oxidation of methane in soil removes about 5.2% of methane from the atmosphere.

Methanotrophic bacteria are widespread in nature. They inhabit all upland, semi-aquatic and aquatic environments of different climate zones. Methanotrophic bacteria or the effects of their activities were observed in arable soils (Hütsch, 1998, Arif et al. 1996, Knief et al. 2005), forest soils (Reay et al. 2001, Wang & Ineson, 2003, Knief et al. 2005), meadow soils (Bender & Conrad, 1993, Horz et al. 2002, Abell et al. 2009), peat soils (Sundh et al. 1995; Mac Donald et al. 1996), rice paddy soils (Kolb et al. 2003, Macalady et al. 2002) in lake sediments (Auman et al. 2000, Costello et al. 2002) and sea sediments (Yan et al. 2006). Moreover, methanotrophs frequently occur in landfill cover soils (Wise et al. 1999, Kallistova et al. 2007, Cebron et al. 2007). They were also isolated from geothermal waters, hot springs (Tsubota et al. 2005, Dunfield et al. 2007), and soda lakes (Khmelenina et al. 1997).

Methanotrophs constitute a numerous and diverse; in morphological, phylogenetic and functional terms; group of Gram-negative bacteria, which still is not fully recognized. In a well-known classification, they are classified as γ or α subdivisions of the Proteobacteria. The α-Proteobacteria consist of *Methylocystis* and *Methylosinus* genera belonging to the family *Methylocystaceae* (McDonald et al. 2008), called as the type II methanotrophs, and two acidophilic genera *Methylocella, Methylocapsa* belonging to the family *Beijerinkiaceae* (Kallistova et al. 2007). The γ-Proteobacteria are represented by the following genera: *Methylobacter, Methylomicrobium, Methylomonas, Methylocaldum, Methylosphaera, Methylothermus, Methylosarcina, Methylohalobius, Methylosoma, Methylococcus* belonging to *Methylococcaceae* family (McDonald et al. 2008). They are classified as the type I methanotrophs. Due to metabolism, type X (also known as Ib) was distinguished within this type. The type X is represented by the genera *Methylococcus* and *Methylocaldum* (Bowman, 2006), which use ribulose monophosphate pathway to the assimilation of formaldehyde. They have also enzymes that are characteristic for serine pathway, typical for the type II methanotrophs (Chistoserdova et al. 2005; McDonald et al. 2008).

In recent years, a novel bacterial strains capable to oxidise methane, belonging to the phylum Verrucomicrobia, has been discovered in the environments with extreme pH, salinity and temperature. For instance, their occurrence has been found in Hell's Gate geothermal area in New Zealand—*Methylacidiphilum infernorum* (earlier known as *Methylokorus infernorum)* (Dunfield et al. 2007); in hot spring of the Kamchatka Peninsula—*Methyloacida kamchatkensis* (Islam et al. 2008) and in Solfatara volcano mud pot in South Italy—*Acidimethylosilex fumarolicum* (Pol et al. 2007).

Until recently, methanotrophs were considered to be obligatory methylotrophic enabled to grow on one-carbon (C_1) compounds but unable to grow on compounds containing C-C bonds. But this statement has been revised. Some members of the genera *Methylocella, Methylocystis,* and *Methylocapsa* are now known to be facultative methanotrophs, capable of growing both on methane as well as on some multicarbon substrates (Dedysh & Dunfiled, 2011). *Methylocella silvestri* originally isolated from an acidic forest soil in Germany can be given as an example. It grows on methane, other C_1 substrates, and on some compounds containing carbon-carbon bonds, such as acetate, pyruvate, propane, and succinate (Chen et al. 2010). The discoveries of recent years provide updated information indicating that several species are characterized by a more flexible metabolism than it was previously assumed (Murell, 2010).

Methanotrophic bacteria conquered many environments that significantly differ between each other in terms of conditions prevailing inside them. It is possible because these bacteria constitute a very large group of microorganisms of different preferences in relation to the availability of substrates (methane and oxygen), temperature of the environment, pH, demand for nutrients, resistance to inhibitors, etc. Most methanotrophs belongs to obligatory mesophilic microerophiles, which prefer the environment with pH close to neutral (Semrau et al. 2010). However, there are many species that prefer conditions radically different from those considered optimal. They belong to extremophilic microorganisms that are able to thrive or prefer to live in extreme habitats or to extemotolerant microorganisms, which are able to withstand or survive in extreme habitats. Methanotrophs able to grow under conditions of high or low pH, temperature or salinity were also found (Trotsenko & Khmelenina, 2002).

An important factor in determining the rate of growth of methanotrophs is the temperature. Genera *Methylobacter* and *Methylomonas* of the type I, and *Methylosinus* and *Methylocystis* belonging to the type II exhibit optimum growth in temperatures in the range of 25 to 35°C. *Methylococcus* (type X), *Methylothermus* and *Methylocaldum* (type I) are thermophiles with an optimum growth ranges between 42°C and 62°C (Tsubota et al. 2005; Halet et al. 2006; Trotsenko & Khmelenina, 2002). Verrucomicrobia belong to thermophiles, whose optimum growth takes place between 55°C to 60°C (Dunfield et al. 2007; Islam et al. 2008). *Methylomonas, Methylobacter* and *Methylosphera;* which belong to the type I; are the psychrophiles that grow in the most effective way at temperatures from 5°C to 15°C (Trotsenko & Khmelenina, 2002). However, the majority of methanotrophs has the ability to adapt to temperatures ranging from 0°C to 55°C (Humer & Lechner, 1999). According to the results of research carried out by Wang et al. (2008), adaptability depends on the type of bacteria. The authors revealed experimentally that in the range of 5°C to 23°C, the number of methanotrophs of the type II in biofilter with compost-peat bed remained stable. Whereas, the number of methanotrophs of the type I clearly decreased with the increase of temperature from 15°C to 23°C and increased with the drop of temperature from 15°C to 5°C .

The majority of methanotrophs grows in neutral environment but many species can tolerate pH in the range of 5.5–8.5 (Hanson & Hanson, 1996). Among methanotrophs, there are also acidophiles of the genera *Methylosinus* and *Methylocystis*, which exhibit optimum growth at pH 5–5.5, and alkaliphiles—*Methylomicrobia*, developing preferably at pH

7.5–10 (Trotsenko & Khmelenina, 2002). What is more, methanotrophic bacteria belonging to Verrucomicrobia; which oxidised methane at pH 3.5 (Islam et al. 2008), at pH of 2–2.5 (Dunfield et al. 2007), and even at pH < 1 (Pol et al. 2007) were also isolated. Most of the methanotrophs prefer the environment of low salinity (Hanson & Hanson, 1996). However, the methanotrophic bacteria (*Methylobacter alcaliphilus* 2OZ, *Methylobacter modestohalophilus* 10S), which grew in the environment of NaCl concentration equalled to 8–9% (Kalyuzhnaya et al. 1999, Khmelenina et al. 1997) were found.

Methanotrophs are characterized by different sensitivity to the concentration of copper in the bed. High concentrations of copper do not favour the development of the type II methanotrophs. It is related to the inhibitory effects of Cu^{2+} ions to the synthesis of sMMO. It was observed that when the concentration of Cu^{2+} exceeds 1 μM, the sMMO synthesis is inhibited. The increase of the concentration of Cu^{2+} ions from 1 to 5 μM stimulates the pMMO synthesis (Hanson & Hanson, 1996). Studies have shown that pure cultures of *Methylococcus capsulatus* (type I) tolerated the Cu^{2+} ions concentration of 30–35 μM, and inhibition followed after the increase in the concentration up to 50 μM (Yu et al. 2003). Moreover, heavy metals such as Hg, Zn and Cr exhibit inhibitory effects on the oxidation of methane. The CH_4 oxidising capacity of arable soils decreased several times after the addition of Hg^{2+} to the concentration of 50 and 100 μg g^{-1} (Contin et al. 2012). It was also noticed that inhibition effect of Zn depended on water regime of soil. Inhibition of CH_4 oxidation in two types of rice soils was observed only under flooded conditions (Mohanty et al. 2000). In contrary, the significant inhibition effect of Cr was observed at 60% moisture holding capacity. The strong inhibitory effect of oxidation of methane is showed by: acetylene (C_2H_2) (Prior & Dalton, 1985), fluoromethane (CH_3F) (Oremland & Culbertson, 1992) and difluoromethane (CH_2F_2) (Miller et al. 1998). Acetylene is a selective MMO inhibitor (DiSpirito, 1991/1992) that binds to the enzyme structure (Urmann et al. 2008). Acetylene inhibits CH_4 oxidation at concentrations as low as 10 μl dm^{-3} while CH_3F and CH_2F_2 inhibited CH_4 oxidation at higher concentrations, equal to 100–1,000 μl dm^{-3} (Chan & Parkin 2000) and 300–500 μl dm^{-3} (Miller et al. 1998), respectively. Furthermore, dichloromethane (DCM) and trichloroethylene (TCE) constitute a competition for the active sites of the monooxygenase. The first one, whose presence completely inhibited methane oxidation in soil, works more effectively (Chiemchaisri et al. 2001). Furthermore, Chiemchaisri et al. (2001) observed the decrease of methane oxidation rate in the presence of tetrachloroethylene (PCE) and benzene, but in the case of those compounds, the inhibition effect was attributed to their toxicity for methane-oxidising bacteria.

Availability of methane determines quality of methanotrophic community. Soils incubated in the atmosphere enriched with methane (10% vol.) showed high capacity for methane oxidation but they do not show the ability to oxidize methane at atmospheric (ambient) concentration (Walkiewicz et al. 2012). This phenomenon can result from the presence of the type II methanotrophs in soils. Those methanotrophs have methanotrophic maximum activity V_{max} and a low affinity to CH_4 (high value of the Michaelis constant K_M) (Bender & Conrad, 1993). Both the parameters are determined on the base of Michaelis-Menten equation. K_M constant represents the substrate concentration at which the rate of an enzyme-catalysed reaction is half of the maximum value. The type II bacteria are isolated from the environments rich in methane (>1%) and poor in oxygen (about 1%) (Hanson & Hanson, 1996), such as bog peat or landfill covers. They are less sensitive to the environmental changes than the type I bacteria which has a low value of V_{max} and high affinity to methane (Henckel et al. 2000) proved by the low value of Michaelis constant (K_M). Species belonging to the type I prefer low concentrations of CH_4 (at atmospheric level). Therefore, they grow mainly on exposed soils at a concentration of methane <1000 ppmv and high concentration of oxygen (Hanson & Hanson, 1996).

5.3.2   *Pathway of aerobic methane biooxidation*

Methane biodegradation is an exoergic process that occurs in the presence of molecular oxygen. The simplified notation of methane oxidation reaction is the following:

$$CH_4 + 2O_2 \rightarrow CO_2 + 2H_2O + energy$$

It is a multistage process which creates a series of intermediates, such as methanol, formaldehyde (methanal) and methanoic (formic) acid. It works only in bacterial cells with the presence of the multi-component enzyme complex, known as methane monooxygenase (MMO). MMO exists in two forms: pMMO (particulate methane monooxygenase) associated with the cell membrane, and sMMO (soluble methane monooxygenase) dissolved in cytoplasm. The presence of pMMO has been reported in all methanotrophs except for the genus *Methylocella* (Theisen et al. 2005); whereas, the sMMO is present only in certain methanotrophs (Murell et al. 2000). The activity of pMMO depends on the presence of copper in the bed material (Buchholz et al. 1995). This type of monooxygenase is inactive in high concentrations of oxygen and access of light (Gilbert et al. 2000).

Methane monooxygenase catalyses the first stage of the process that is the oxidation of methane to methanol. In the next step of the process, involving the use of methanol dehydrogenase, formaldehyde is formed. This substance is later oxidised to formic acid and ultimately to carbon dioxide with the use of formate dehydrogenase (Fig. 5.3). Part of the carbon contained in formic aldehyde may be incorporated into the bacterial cell material in ribulose monophosphate cycle (RuMP), which occurs in the type I methanotrophs or serine cycle occurring in the type II methanotrophs (Hanson & Hanson, 1996). Therefore, the fate of carbon particle originating from methane is the basis for the separation of two types of the pathways: dissimilative, whereby carbon is reincorporated into CO_2 and released to the atmosphere and is not utilized in cell biomass, and the assimilative, whereby the carbon from CH_4 is built up into organic compounds forming cellular biomass.

The process of methane oxidation is exoenergetic, and the largest portion of energy is released during the stages 3 and 4 of the process, in which formaldehyde is transformed into methanoic acid (formate), and from that into inorganic end products.

Stoichiometrically, 1 mole of CO_2 is produced as a result of oxidation of 1 mole of CH_4. However, in practice, a part of carbon is assimilated by bacteria, whereby a proportion of generated CO_2 to consumed CH_4 is lower than 1. The value of the proportion depends on the level of synthesis of new cell material (Hoeks, 1972) that, in turn, depends on the phase of the growth of microorganisms population. In still growing, young bacteria community

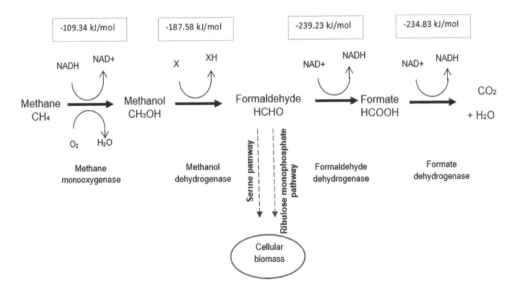

Figure 5.3　Scheme of CH_4 oxidation pathway (according to Hoeks, 1972 and Brigmon 2001, modified).

the assimilative pathway prevails over dissimilative, reducing the emission of CO_2 in relation to the value resulting from the stoichiometry of the reaction. With the population aging less carbon is assimilated, whereby CH_4/CO_2 ratio approaches the theoretical value.

5.3.3   *Methanotrophs in landfill covers and biofilters*

Methanotrophs isolated from landfills covers, methanotrophic biofilters and model laboratory systems, through which methane was passed, belonged to both the type II, which prefers high concentrations of CH_4, as well as the type I. The most commonly isolated type II bacteria were the following: *Methylocystis* (Nozhevnikova et al. 1993, Nikiema et al. 2005, Stralis-Pavese et al. 2004, Hilger et al. 2007, Cebron et al. 2007, Chen et al. 2007). Whereas, the representatives of genus: *Methylocella* (Chen et al. 2007, Cebron et al. 2007), *Methylosinus* (Nozhevnikova et al. 1993) and *Methylocapsa* (Cebron et al. 2007) were rarely found. The type I was the most represented by bacteria such as *Methylobacter* and *Methylomonas* (Nozhevnikova et al. 1993; Cebron et al. 2007, Chen et al. 2007, Hilger et al. 2007, Jugnia et al. 2009). However, the presence of *Methylosarcina* (Chen et al. 2007), *Methylocaldum* (Stralis-Pavese et al. 2004), and *Methylococcus* (Gebert et al. 2009) was also noted (Table 5.2).

When it comes to the quantitative proportions between microorganisms of type I and type II, there is no clear dominance of one over another. Wang et al. (2008) showed that in the laboratory biofilter filled with a mixture of peat and compost from sewage sludge, the number of the type I methanotrophs was 5-fold higher than type II bacteria. The type I methanotrophs prevailed also in the soil from landfill cover examined by Kallistova et al. (2007). In the laboratory biofilter filled with clay soil, the ratio of the number of type I to the type II bacteria ranged from 9.16 to 13.93 (He et al. 2008). However, when a fine fraction obtained after sieving anaerobically stabilized waste was examined in lysimeters the dominance of the type II methanotrophs or a comparable number of both types (the ratio of the number of bacteria of type I to type II ranged from 0.3 to 1.2) were found. The type II was also numerously represented in methanotrophic biofilters investigated on a laboratory

Table 5.2   Methanotrophs isolated from landfill covers or methanotrophic biofilters.

γ-Proteobacteria	α-Proteobacteria	References
Methylobacter *Methylomonas*	*Methylocystis* *Methylocella* *Methylocapsa*	Cebron et al. (2007)
Methylobacter *Methylomonas* *Methylosarcina*	*Methylocystis* *Methylocella*	Chen et al. (2007)
Methylobacter *Methylococcus*	*Methylocystis*	Gebert et al. (2009)
Methylobacter *Methylobacter* *Methylosoma* *Methylocaldum* *Methylococcus*	*Methylocystis* *Methylocystis*	Hilger et al. (2007) Jugnia et al. (2009) Kong et al. (2013)
– *Methylobacter bovis, Methylobacter hroococcum Methylobacter capsulatus Methylomonas albus*	*Methylocystis parvus Methylocystis minimus Methylocystis pyriformis Methylocystis echimoides Methylocystis parvus Methylosinus sporium Methylosinus trichosporium*	Nikiema et al. (2005) Nozhevnikova et al. (1993)
Methylocaldum	*Methylocystis*	Stralis-Pavese et al. (2004)

Table 5.3   Number of methanotrophs in landfill cover soils and LFG biofilters.

Source of soil sample	Bacteria number [cells g^{-1} d.w]	References
Landfill cover	1.5×10^9	Ait-Benichou et al. (2009)
Field-scale biofilter filled with compost	2×10^{10}–2×10^{11}*	Dammann et al. (1999)
Field-scale biofilter filled with expanded clay pellets	1.32×10^8–1.22×10^{11}	Gebert et al. (2003)
Field-scale biofilter	3×10^9–5×10^{11}	Gebert et al. (2008)
Field-scale biofilter	4×10^8–1×10^{10}	
Landfill cover	$15 \pm 2 \times 10^8$–$56 \pm 7 \times 10^8$	Kallistova et al. (2007)
Laboratory biofilter filled with soil from landfill cover	4.0×10^7	Park et al. (2008)
Landfill biocover	3.91×10^8	Cabral et al. (2010)
Laboratory biofilter filled with		Pawłowska et al. (2011)
– horticultural substrate	7.39×10^4–6.38×10^5	
– municipal waste compost	2.49×10^6–7.85×10^6	
– mixture based of perlite and compost	3.20×10^6–1.81×10^7	

*cells per g.

scale by Stralis-Pavese et al. (2004). The discrepancies of observations can be explained by the different conditions under which methanotrophs developed, resulting from the difference of the pH value of the bed material or the availability of substrates. Kong et al. (2013) observed that compared to the type I methanotrophs, the type II methanotrophs (belonging to *Methylocystis*), were more abundant in the acidic landfill cover soils. It was also observed that the type I methanotrophs occurs more frequently in the surface parts of the biofilter or landfill covers, where the oxygen concentration is high and the concentration of methane relatively small. High concentrations of methane in landfill gas and oxygen deficiency lead to the domination of the type II bacteria in the composition of the population (Stralis-Pavese et al. 2004; Wang et al. 2008). What is more, the concentration of copper ions in the bed material may be another cause of the differences in observation results. The sMMO, which is produced by the type II and the type X methanotrophs, is expressed only under low copper concentration in soil (Grosse et al. 1999).

In biofilters and landfill covers, the number of methanotrophs, calculated per gram of dry mass of the filter material, is typically from 10^7 to 10^{11} (Table 5.3). It is relatively high in comparison to other environments exposed to high concentrations of methane. For instance, the number of methanotrophs in tropical dry land rice field soil was estimated to 5.29–73.66×10^6 cells g^{-1} dry weight (Dubey & Singh, 2000), and in fresh water lake sediment to 3.6–7.4×10^8 cells g^{-1} dry weight (Costello et al. 2000).

The population of methanotrophs in landfill covers is approximately 50% of the total number of bacteria (Kallistova et al. 2007). The participation of methanotrophs in methanotrophic biofilters may be even higher. According to the study conducted by Wang et al. (2008), their participation oscillated between 51.5 to 67.3% of the total number of bacteria.

5.4   BIOOXIDATION OF VOCs UNDER AEROBIC CONDITIONS

Many organic compounds such as aliphatic hydrocarbons, aromatic hydrocarbons and their derivatives, which are present in the biogas; can be biodegradable in soil and water. These compounds can be decomposed, constituting a primary substrate used as a sole source of carbon and energy or as a cosubstrate, not assimilated by microorganisms. The result of biodegradation of the compound, used as a primary substrate, is complete degradation (mineralization) of organic pollutants and the microorganism cell growth. In the second case, the transformation process of the compound proceeds *via* cometabolism, which is defined as the metabolism of an organic compound without nutritional benefit for microorganism occurring in the presence of a growth substrate. The cosubstrate is not assimilated, but the product of its disintegration may be available as substrate for other organisms of a mixed culture (Fritsche & Hofrichter, 2008).

5.4.1 *VOCs-oxidising microorganisms*

The oxidation of volatile organic compounds is performed by many prokaryotic and eukaryotic microorganisms inhabiting soil and water: *Pseudomonas, Xanthobacter, Sphingomonas, Burkholderia, Alcaligenes, Hypromicrobium, Methylobacterium, Methylosinus, and Paracoccus* (Table 5.4). *Pseudomonas* is the most common genus of bacteria able to adapt to diverse substrates. Bacteria belonging to this genus possess several catabolic pathways, which allows them to use different substrates, also that recalcitrant to biodegradation, such as aromatic hydrocarbons. The genus gathers over 200 species. One of them is *P. putida* which possesses metabolic pathways dedicated to benzene, toluene and xylenes. But it also can degrade other hydrocarbon, such as p-cymene and trichloroethylene. What is more, Gram-positive bacteria belonging to *Rhodococcus and Arthrobacter* is also capable to degrade volatile organic compounds (Table 5.4).

Capability to degrade of many hydrocarbons both aliphatic and aromatic was found also in the case of some genera of fungi, e.g. *Paecilomyces, Fusarium, Cladosporium, Exophiala, and Phanerochaete* (Table 5.4). Qi et al. (2002) showed that among five fungal species analysed, that is: *Exophiala lecanii-corni, Cladosporium sphaerospermum, Cladosporium resinae, Mucor rouxii,* and *Phanerochaete chrysosporium. E. lecanii-corni* and *C. sphaerospermum* had the broadest spectrum of substrates used as a carbon and energy source. They were able to degrade some organic acids, ketones, and aromatic compounds (Table 5.4). When it comes to *P. chrysosporium*, it could grow on many compounds analysed but it was not able to degrade styrene. *M. rouxii*, which showed visible growth only when supplied with n-butyl acetate or ethyl 3-ethoxypropionate, turned out to be the least useful in biodegradation.

5.4.2 *Pathways of degradation of VOCs used as a primary substrate for bacteria growth*

Process of biooxidation of organic compounds associated with bacteria growth (Fig. 5.4) generally proceeds in the following steps (Fritsche & Hofrichter, 2008):

1. contact between the microbial cells and the organic pollutants particle, and transport of the compounds to the cell,
2. initial intracellular attack on organic pollutants by oxygenases or peroxidases in the presence of molecular oxygen,
3. degradation on pollutant by peripheral pathways—conversion of organic pollutants into intermediates of the central intermediary metabolism, e.g., the tricarboxylic acid cycle,
4. biosynthesis of cell biomass from the central precursor metabolites, e.g., acetyl-CoA, succinate, pyruvate. Sugars required for various biosynthesis and growth must be synthesized by gluconeogenesis.

Aliphatic and cycloaliphatic hydrocarbons may be used as substrates, whose decomposition leads to the growth of cells. The main pathway of biodegradation of alkenes is monoterminal oxidation that proceeds *via* the formation of the corresponding alcohol, aldehyde, and fatty acid. β-Oxidation of the fatty acids leads to formation of acetyl-*Co*A that is incorporated to the intermediary metabolism.

The use of aromatic compounds associated with cell growth is also possible. Most polycyclic and highly polymerized organic compounds may only be decomposed under aerobic conditions. Molecular oxygen is used in the initial attack on an aromatic ring and leads to its disruption. Benzene, chlorinated benzenes, substituted benzenes are biodegraded under aerobic conditions, serving as sole carbon and energy sources for numerous species of bacteria belonging to genera *Pseudomonas, Sphingomonas, Rhodococcus, Burkholderia* and other.

Table 5.4   Some microorganisms capable of degrading VOCs founded in LFG.

Microorganisms	Compound	References
BACTERIA		
Alcaligenes denitryficans	Benzene, toluene, methylbenzene	Ridgway et al. (1990)
Arthrobacter sp.	Dimethylsilanediol (DMSD)	Sabourin et al. (1996)
Burkholderia cepacia	Trichloroethylene	Mars et al. (1996)
	Toluene	Shields & Montgomery (1989)
Hyphomicrobium chloromethanicum	Chloromethane	McDonald et al. (2001)
Methylophilus sp.	Dichloromethane	Bader and Leisinger (1994)
Methylosinus trichosporium	Trichloroethylene	Mars et al. (1996)
Pseudomonas sp.	Chlorobenzen	Kunze et al. (2009)
	1,4-Dichlorobenzene	Sommer & Görisch (1997)
	Styrene	Gąszczak et al. (2009)
	Dichloromethane	Guo et al. (1990)
Pseudomonas putida	p-Cymene	Eaton (1997)
	Trichloroethylene	Mars et al. (1996)
	Benzene, ethyl benzene, toluene, m-xylene, p-xylene, o-xylene	Gülensoy & Alvarez (1999)
Pseudomonas aeruginosa	Benzene	Kim et al. (2003)
Pseudomonas stutzeri	o-Xylene	Barbieri et al. (1993)
Pseudomonas veroni	Alkyl methyl ketones, 2-butanol, 2-hexanol	Onaca et al. (2007)
Paracoccus versutus	Acetone	Su et al. (2012)
Ralstonia pickettii	Chlorobenzen	Zhang et al. (2011)
Rhodococcus sp.	Benzene, phenol, toluene, ethyl benzene, isopropyl benzene,	Kim et al. (2002)
Sphingomonas haloaromaticamans	1,4-Dichlorobenzene	Sommer & Görisch (1997)
Xanthobacter sp.	Dichloromethane	Emanuelson et al. (2009)
Xanthobacter flavus	1,4-Dichlorobenzene	Sommer & Görisch (1997)
FUNGI		
Cladosporium resinae	Ethyl benzene and toluene, methyl ethyl ketone, methyl isobutyl ketone	Qi et al. (2002)
Cladosporium sphaerospermum	BTEX, styrene, methyl ethyl ketone, methyl isobutyl ketone	Qi et al. (2002)
Exophiala lecanii-corni	BTEX, styrene, methyl ethyl ketone, methyl isobutyl ketone	Qi et al. (2002)
Fusarium oxysporum	Dimethylsilanediol (DMSD)	Sabourin et al. (1996)
Phanerochaete chrysosporium	BTEX, methyl ethyl ketone, methyl isobutyl ketone	Qi et al. (2002)

Due to the fact that aromatic compounds are characterized by a higher thermodynamic stability than aliphatics, their decomposition process is more complex. The main role in this process is played by oxygenase enzymes that are able to cleave the aromatic ring of the compound. Aerobic degradation of hydrophobic pollutants such as benzene, toluene, naphthalene, biphenyl or polycyclic aromatics is usually initiated by activation of the aromatic ring through oxygenation, catalysed by members of the soluble di-iron family of monooxygenases (Leahy et al. 2003) or Rieske non-haem iron oxygenases (Gibson & Parales, 2000). The possibility of bacteria to degrade many aromatic hydrocarbons is based on the existence inside the cell of catabolic plasmids (e.g. TOL plasmid) that encode specific enzymes. The many of aromatics can be converted to intermediates, such as catechol, 3-methyl catechol or protecatechuate (Fritsche & Hofrichter, 2008, Gülensoy & Alvarez, 1999).

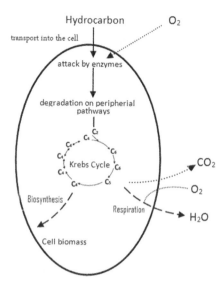

Figure 5.4   Scheme of aerobic biodegradation of hydrocarbons associated with growth of microor-ganism (Fritsche & Hofrichter, 2008).

The pathway of aerobic toluene biodegradation has been widely studied and can serve as a model for understanding the mechanisms of bacterial metabolism of benzene ring compounds. *Pseudomonas* strains can degrade toluene by different oxygen-dependent pathways, named after the operons that code them (Zylstra, 1994). According to the data collected by Fang et al. (2000) the following pathways are known:

1. TOL pathway, discovered in *P. putida (arvilla)* mt-2, is coded in the extra chromosomal DNA, called tol pWW0 plasmid. In the initial degradation steps a methyl group is oxidized by a monooxygenase to the benzyl alcohol (Fig. 5.5), which is converted to catechol in the next steps. TOL plasmids in *P. putida* encode the metabolic pathways also for the degradation of xylenes, as well as alcohol and carboxylate derivatives of toluene and xylenes (Assinder & Williams, 1990).
2. TOD pathway occurs in *P. putida* F1, due to the catalytic activities of the enzymes encoded by the chromosomally located *tod* genes (Zylstra et al. 1988). Toluene dioxygenase is used to convert toluene into *cis*-toluene dihydrodiol (Gibson et al. 1968), which is subsequently dehydrogenated to 3-methylcatechol (Fig. 5.5). Due to low substrate specificity of toluene dioxygenase this pathway leads to the oxidation of more than 100 substrates (Gibson et al. 1995).
3. TOM pathway occurs in *Burkholderia cepacia* G4 (formerly known as *Pseudomonas cepacia* G4), which uses toluene *ortho*-monooxygenase in the initial attack to form *o*-cresol (Shields & Montgomery, 1989), which is converted to 3-methylcatechol (Fig. 5.5).
4. TBU pathway occurs in *Burkholderia pickettii* PK01 (formerly known as *Pseudomonas pickettii* PK01), which hydroxylates the aromatic ring using toluene *meta*-monooxygenase to produce *m*-cresol (Olson et al, 1994), which is transformed to 3-methylcatechol (Fig. 5.5).
5. T4MO pathway expressed by *P. mendocina* KR1, which uses toluene para-monooxygenase to produce *p*-cresol (Whited & Gibson, 1991), which is converted to protecatechuic acid (Fig. 5.5).

In the next metabolic steps the aromatic ring is cleaved and the compounds are converted *via meta* or *ortho* fission to pyruvate or succinate and acetyl-Co-A (Fig. 5.6). There are the key elements of intermediary metabolic pathways, necessary for the synthesis of cell building blocks or energy production.

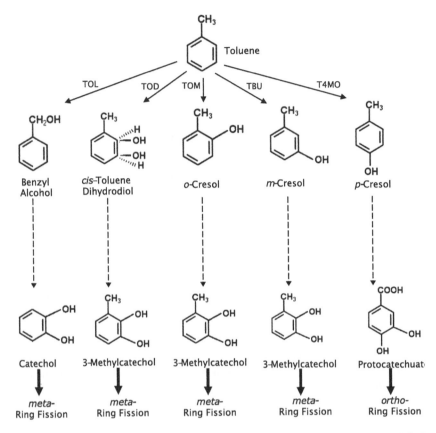

Figure 5.5   Aerobic biodegradation pathways of toluene (Gülensoy & Alvarez, 1999, modified).

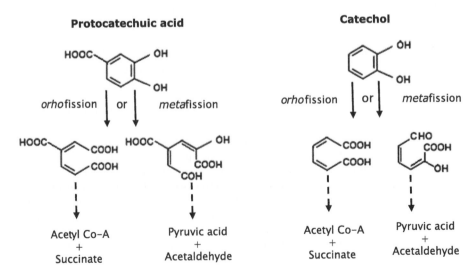

Figure 5.6   *Ortho* and *meta* fission of aromatic ring of protocatechuic acid and catechol (Juhasz & Naidu, 2000, Guzik et al. 2013).

5.4.3   *Cometabolic pathways of aerobic VOCs biodegradation*

The degradation of many hydrocarbons, even chlorinated, may also happen during cometabolism of other compounds. For instance; the substances such as methane, propane, toluene, cumene, limonene are used as the basic substrates required for the growth of microorganisms (Bioaugmentation for Remediation, 2005). Methanotrophs, using methane as a carbon and energy source, are a well-known group of microorganisms having the ability to cometabolic decomposition of many VOCs. Methanotrophs are considered to be important organisms, which can transform some complex organic compounds to susceptible substrate, usually simply hydroxylated derivatives, easily utilized by heterotrophs (Hršak & Begonja, 2000).

There are many reports about cometabolic activity of methanotrophs in the transformation of multicarbon compounds, also chlorinated. Colby et al. (1977) stated that sMMO of *Methylococcus capsulatus* can hydroxylate C1-C8 n-alkanes yielding mixtures of 1- and 2-alcohols. Those are also hydroxylates cyclic alkanes and aromatic compounds. However, styrene yields only styrene epoxide and pyridine yields only pyridine N-oxide. Many researchers have studied capability of methane-grown bacteria to the biodegradation of chlorinated compounds. That is well recognized that many of chlorinated aliphatic compounds can be degraded by a diversity of bacteria including methanotrophs (Little et al. 1988). One of the most frequently studied compound is trichloroethylene (TCE). In the 1980s, Fogel et al. (1986) claimed that TCE and other chlorinated compounds, such as vinyl chloride and vinylidene chloride, were degraded by the mixture culture of microorganisms obtained by methane enrichment of sediment samples to the products, which were not included in volatile chlorinated substances. Two other chlorinated ethenes, *cis* and *trans*-1,2-dichloroethylene, were degraded to chlorinated products. Whereas, the capability of methanotrophs to the degradation of tetrachloroethylene (PCE) was not confirmed (Fogel et al. 1986, Bowman et al. 1993).

Researchers noticed that the biodegradation of TCE was inhibited by acetylene, a specific inhibitor of MMO. This observation supported the hypothesis that methanotrophs were responsible for TCE degradation. The study conducted by Little et al. (1988) confirmed this hypothesis and identified a possible course of TCE decomposition. The pathway of the cometabolic degradation of TCE is presented in Figure 5.7. McCarty & Semprini (1994) noticed

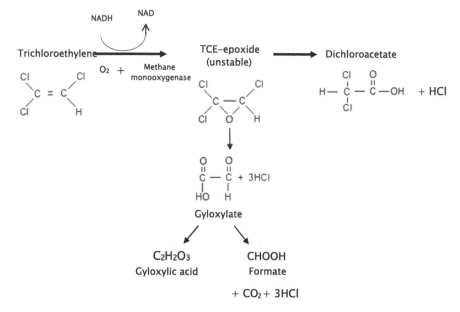

Figure 5.7   Pathway of TCE biodegradation by methanotrophs (Brigmon, 2001).

that biodegradation of TCE was strongly tied to methane utilization. Whenever the methane addition were withheld, TCE transformation stopped. Frascari et al. (2006) revealed that aerobic bacteria growing on methane and propane have the ability to oxidize 1,1,2,2-tetrachloroethane (TeCA). Hesselsoe et al. (2005) showed the degradation of vinyl chloride by pure culture of the methanotroph *Methylosinus trichosporium* OB3b. They also stated that tested strain was able to oxidize benzene but the process was ceased after only 2 days. Degradation of benzene by enriched methane-oxidizing consortium isolated from landfill cover soil was observed by Lee et al. (2011a). They observed that methane oxidation rates decreased with the increase in benzene/methane ratio and benzene oxidation rates increased with the increase in the ratio. Lee et al. (2010) stated that benzene and toluene oxidation in landfill cover soil was stimulated by the presence of methane. According to the results of the study carried by Gebert et al. (2008), it is suggested that bacteria belonging to *Methylosinus* species play an important role in the process of co-oxidative degradation of non-methane VOCs. The authors found that the community of the type II methanotrophs in biofilter purged with the model landfill gas enriched with NMVOCs was dominated by *Methylosinus* sp.; whereas the biofilter purged with the gas, which consisted only of CH_4, CO_2, N_2, was dominated by *Methylocystis* sp.

Cometabolism observed at methanotrophs is a result of nonspecific methane monooxygenase activity towards organic compounds that do not serve as carbon or energy sources (Brigmon, 2001). A soluble form of monooxygenase (sMMO), which is present in the type II methanotrophs, has low substrate specificity. It is able to oxidize several alkanes and alkenes, cyclic hydrocarbons, aromatics, and halogenic aromatics (Grosse et al. 1999); for instance, trans-dichloroethylene, vinyl chloride (Yoon & Semrau, 2008), dichloromethane (DCM) (Chiemchaisri et al. 2001), and 1,1,1-trichloroethane (TCA) (Kjeldsen et al. 1997).

Studies indicate that the membrane-associated particulate methane monooxygenase (pMMO), regarded as having a high substrate specificity, can also catalyse the decomposition of hydrocarbons other than methane. DiSpirito et al. (1991/1992) observed degradation of trichloroethylene (TCE) by pMMO expressing bacteria, but the oxidation rate of TCE was lower than those catalysed by the sMMO. Yoon & Semrau (2008) suggested that type of monooxygenase involved in the cometabolic decomposition of chlorinated ethenes depends on the ambient temperature, which may exert selective influence on methanotrophic communities to express sMMO or pMMO. When *Methylosinus trichosporium* OB3b was grown at 30°C, greater degradation of chlorinated ethenes was observed under particulate methane monooxygenase (pMMO)-expressing conditions than sMMO-expressing conditions; while at 20°C the maximal rates of chlorinated ethane degradation were greater by *M. trichosporium* OB3b expressing sMMO as compared with the same cell expressing pMMO.

Cometabolic conversion of chlorinated ethenes has been described also for several types of non-methanotrophic microorganisms, including species of *Pseudomonas* and *Burkholderia* capable of degrading aromatic compounds. Ensley (1991) has demonstrated the possibility of TCE degradation by *P. cepacia* G4 (now called *Burkholderia cepacia*), *P. mendocina* and *P. putida* which can grow on aromatic compounds. Beforehand, Nelson et al. (1987) had shown that the ability of TCE degradation by *Burkholderia cepacia* G4 was possible when grown on toluene or other aromatic compounds as carbon and energy source. This is possible due to low substrate specificity of toluene-monooxygenase. According to Sazinsky (2004) the toluene/o-xylene monooxygenase (ToMO) from *Pseudomonas stutzeri* OX1 is capable of oxidizing arenes, alkenes, and haloalkanes at a carboxylate-bridged diiron center similar to that of soluble methane monooxygenase (sMMO). Easign et al. (1992) reported that pure cultures of *Xanthobacter* spp. cometabolised TCE with the utilization of propylene as a carbon and energy source presumably using the enzyme alkene monooxygenase.

5.4.4 *Substrate interactions affecting biodegradation of particular BTEXs*

Benzene, ethyl benzene, toluene, and xylenes are the most common aromatic hydrocarbons in LFG. Many species of bacteria are able to degrade all BTEX or some of them (Table 5.5).

Table 5.5 Catabolic profiles of different microorganisms responsible for biodegradation of BTEX in aerobic conditions.

Microorganism	Benzene	Toluene	Ethyl-benzene	p-Xylene	m-Xylene	o-Xylene	References
Alcaligenes denitryficans (16 strains)	+	+	+	0	n.a	0	Ridgway et al. (1990)
Burkholderia cepacia G4	+	+	+	+	+	+	Gülensoy & Alvarez (1999), Duetz et al. (1994), Marsa et al. (1996)
Burkholderia cepacia MB2	n.a.	n.a.	n.a.	n.a.	n.a.	+	Jørgensen et al. (1995)
Burkholderia pickettii PK01	0	0	+	+	0	+	Gülensoy & Alvarez (1999)
Clavibacter michiganense Bp11	0	0	0	+	0	0	Gülensoy & Alvarez (1999)
Comamonas testosterone 27911	+	+	0	0	0	0	Gülensoy & Alvarez (1999)
Nocardia spp.	+	+	+	0	0	+	Ridgway et al. (1990)
Pseudomonas alcaligines (2 strains)	+	+	+	0	n.a.	0	Ridgway et al. (1990)
Ps. aeruginosa N4	0	+	+	+	+	+	Gülensoy & Alvarez (1999)
Ps. aeruginosa (13 strains)	+	+	+ 11 strains	+ 11 strains	+	+ 1 strain	Ridgway et al. (1990)
Ps. corrugata NG2	0	+	+	+	+	+	Gülensoy & Alvarez (1999)
Ps. mendocina KR1	+	+	+	+	+	0	Gülensoy & Alvarez (1999), Yen et al. (1991), Duetz et al. (1994)
Ps. putida CCMI 852 with a TOL plasmid	0	+	n.a	+	+	+	Otenio et al. (2005)
Ps. putida F1	+	+	+	+	+	+	Gülensoy & Alvarez (1999), Duetz et al. (1994), Eaton (1997)
Ps. putida mt-2	0	+	+	+	+	0	Duetz et al. (1994), Gülensoy & Alvarez (1999)
Pseudomonas fluorescens Type F	0	+	+	+	+	+	Gülensoy & Alvarez (1999)
Pseudoxanthomonas spadix BD-a59	+	+	+	+	+	+	Choi et al. (2013)
Pseudomonas sp. CFS-215	+	+	+	+	+	+	Gülensoy & Alvarez (1999)
Rhodococcus spp.				+		+	Jang et al. (2005)
fungus Paecilomyces variotii CBS115145	+	+	+	+	+	+	Garcia-Pena et al. (2008)

+ Degraded, 0 Not degraded, n.a. Not analysed.

They can be utilized both as a substrate for bacteria growth and as a co-substrate, not using as carbon and energy source.

BTEX degradation was evaluated both in the mixture cultures inhabiting soils or sediments and in the pure culture of microorganisms. The screening of 297 gasoline-degrading bacteria belonging to genera *Pseudomonas*, *Alcaligenes*, *Nocardia*, and *Micrococcus*, isolated from well water and core material from a shallow coastal aquifer contaminated with unleaded gasoline revealed their ability to degrade many hydrocarbons, among them BTEX. Most isolates were able to grow on 2 or 3 different hydrocarbons, and nearly 75% utilized toluene as a sole source of carbon and energy. Toluene, p-xylene, ethyl benzene were the most frequently utilized as growth substrates, while benzene was the one of the least utilized compounds (Ridgway et al. 1990). In accordance with the research, the different possibilities of using various substrates might concern even only one species of bacteria. It was shown that *Pseudomonas putida* may utilized aromatic hydrocarbon on two different pathways: TOD and TOL. *P. putida F1* using TOD pathway can metabolize all BTEX (Gülensoy & Alvarez, 1999; Choi et al. 2003). But *P. putida (arvilla)* mt-2 using the TOL pathway is able to degrade toluene, methylbenzene and *m*- and *p*-xylenes, but it cannot utilise benzene and *o*-xylene (Gülensoy & Alvarez, 1999). Also *P. putida* CCMI 852 with a TOL plasmid, examined by Otenio et al. (2005), did not have the ability to degrade benzene. It was not metabolized both as a separate element, and as a component of the mixture with toluene and xylenes.

The possibility to utilise different substrates causes that the biodegradation of BTEX is affected by substrate interactions. Arvin et al. (1989) observed an antagonist effect of each toluene and xylenes when they were present together in headspace over the bacteria colony. The rates of their degradation were lower than toluene and xylene alone. The antagonist effect of these compounds was observed also by Otenio et al. (2005). The degradation rates of toluene and xylene in the mixture of BTX (with the equal participation of all components) decreased by 50%, in comparison to the rates of these compounds removal in the one-substrate tests. Research carried out by Gülensoy & Alvarez (1999) on 55 strains of different genera of bacteria showed a significant correlation between the abilities to degrade following couples of compounds: toluene and ethyl benzene, p-xylene and m-xylene, p-xylene and o-xylene. The study showed that the inability to degrade benzene was correlated to the inability to degrade o-xylene. Alvarez & Vogel (1991) have shown that the presence of toluene in a gas mixture increased the rate of benzene degradation of in pure cultures of *Pseudomonas* sp. However, toluene was also reported to competitively inhibit benzene degradation in experiments with different bacteria strains (Oh et al. 1994). These divergent observations show that it is not possible to draw general conclusions about the capacity of individual BTX compounds to enhance or inhibit the degradation of other BTX compounds (Alvarez & Vogel, 1991). These differences can be explained by the diversity of the composition of the consortium of microorganisms involved in the process.

Substrate interactions was observed also during the BTEX degradation by filamentous fungus *Paecilomyces variotii* (Garcia-Pena et al. 2008). The examination carried out in liquid culture showed that toluene was the most susceptible to degradation. It was completely degraded. Benzene, *m*- and *p*-xylenes was degraded by 45%, while *o*-xylene only by 30%. In mixtures of toluene–benzene the toluene degradation rate was *ca.* 30% lower than the rate obtained with only toluene, while in the mixture of ethyl benzene–benzene the rate for ethylbenzene removal was also *ca.* 30% lower as a single substrate. Benzene degradation was also negatively affected by both toluene and ethyl benzene.

5.5   FACTORS DETERMINING EFFICIENCY OF BIOLOGICAL METHODS FOR MITIGATION OF LFG EMISSION

The effectiveness of biological methods in mitigation of landfill gas depends on the amount and the composition of microorganisms consortia in the biofilter. The development of microorganisms is determined by many factors, which can be classified into three groups: composition and properties of the bed material; composition and amount of purified gas; and external conditions, such as temperature and pressure. As a result of overlapping of the effects of various factors; specific community of microorganisms, which is sensitive to environmental changes, is being created. Knowledge of habitat requirements of specific groups of microorganisms enables to control the direction of the development of community in the bed. It is aimed at obtaining the greatest efficiency of gas removal from the mixture with a specific chemical composition. It is not possible to create conditions that will allow for maximum ability to remove all biodegradable compounds in the bed. Habitat requirements of different groups of microorganisms differ significantly. Therefore, it is necessary to focus on compounds which constitute the most dangerous threat to the environment. Due to the high concentration of methane in the landfill gas, in the case of biofilters used in landfills, the aim is to create the best conditions for the growth of methanotrophic bacteria. An additional argument in favour of such activities is the fact that methanotrophs also possess the capability for cometabolic degradation of many trace compounds contained in the biogas. Thus, the development of such bacteria community allows for the conversion of a number of complex hydrocarbons into simpler compounds, which may become available as substrates for other groups of microorganisms.

5.5.1 *Parameters of filter bed material*

Filter bed material plays the key role in the development of a consortium of microorganisms. The material must be stable in terms of physical and chemical state; have an appropriate granulometric composition, which determines the optimal water and air conditions and a large specific surface area; be rich in macro and micronutrients; have no compounds toxic to bacteria; provide easy access of substrates and discharge of metabolites. The material should be homogeneous in order to avoid zones, which differ significantly according to the growth of microorganisms or flow resistance for gases (Humer-Huber et al. 2008). It should also be resistant to sedimentation.

Granulometric and mineral composition of the material decide about its physical properties. They determine porosity, specific surface area, water capacity, and gas and water permeability. Those properties determine the surface area available for colonization by microorganisms and the conditions of water and air, which are significant due to the availability of water, oxygen and substrates for microbial cells. Porosity determines the properties of water and air by adjusting the gas diffusion profile of the filter bed material. The specific surface area affects the development of a bacterial population (which form a biofilm on the solid phase), sorption properties of the filter bed material, nutrients retention capacity (reducing their vulnerability to leaching) and water; as well as increase of buffering properties with respect to changes in pH. When it comes to the granulometric composition, it is also very important in terms of ensuring the low pressure loss during the passage of gas through the bed.

Regarding the gas flow facility; the coarse mineral materials with significant share of the sand and gravel fractions possess the most suitable conditions for gas migration. However, their disadvantage is a small specific surface area and a lack of nutrients. It was indicated that the mineral materials with medium and coarse sand have a greater ability to oxidize CH_4 than materials with a grain size of gravels (Kightley et al. 1995, Pawłowska et al. 2003).

In order to create good conditions for the development of microorganisms and the flow of gas through the bed, an appropriate balance between porosity and the surface of a material is required. A large total porosity (>60%) and a content of macropores above 25% should be considered optimal conditions (Huber-Humer et al. 2009). Various kinds of compost meet such parameters. Moreover, due to their high content of organic matter, they constitute a rich source of nutrients and have a pH value that is near to neutral; which is preferred for many species of microorganisms responsible for the biodegradation of methane and other compounds present in the biogas.

The importance of organic matter as a component of the filter bed material used in the biofiltration of landfill gas is highlighted by the results of laboratory tests conducted in continuously flow systems with passive aeration (oxygen diffusion through the surface of the bed). Materials rich in organic matter are characterized by methane oxidation capacity that ranges from 160 to 400 g CH_4 $m^{-2}d^{-1}$ (Powelson et al. 2006, Wilshusen et al. 2004b). Whereas, methane oxidation capacity of mineral soils, sands and gravels ranges from 102 to 240 g CH_4 $m^{-2}d^{-1}$ (Kightley et al. 1995, De Visscher et al. 1999, Pawłowska et al. 2003). Christophersen et al. (2000) found that composts and other materials with a concentration of organic matter >35% by weight have up to 10–100 times higher rate of methane oxidation than the materials containing less of their quantity (e.g. mineral soils containing at most a few percent of the organic matter).

However, not all composts are characterized by equal suitability for their usage in the biofiltration of landfill gas, which was stated during the research on the removal of methane (Wilshusen et al. 2004b). The methanotrophic capacity of compost is also influenced by the composition of the raw material, which they originated from; the degree of stabilization of organic matter; as well as composting technology. These factors determine the properties of the compost. The most important physical and chemical properties, which should be taken into account when using the compost in the biofiltration of landfill gas, are shown in Table 5.6. A respiration activity, which indicates the compost maturity, is a significant parameter. Immature compost is less suitable because the rapid mineralization of organic substance, which takes place inside it, may lead to a reduction of the oxygen concentration. Therefore, the biodegradation processes of methane might be inhibited (Lechner & Huber, 1999).

Granulometric composition and organic matter content influence water properties of the material. Water is essential for the life of microorganisms; however, its excess limits the passage of gas to microbial cells, decreasing at the same time their diffusion rate and consequently lowering the efficiency of contaminants removal. In addition, the excess of water can promote the growth of fungi, which interferes with the normal growth of methanotrophs. Depending on the technological solution of biofiltration; atmospheric precipitation, sprinkler systems or gas humidification systems, reactions occurring within the bed and water vapour contained in the gas landfill are the source of water in the bed (water is a by-product of the oxidation of CH_4, the mineralization of organic matter). Organic materials are characterized by high water retention. Therefore, in order to improve air and water conditions;

Table 5.6 Parameters of compost materials used for testing the suitability for biocover construction recommended by Huber-Humer et al. (2009).

Parameter	Recommended value
Bulk density [kg dm^{-3}]	0.8–1.1
Moisture content [% water holding capacity]	50
Water holding capacity [% d.w.]	50–130
Air porosity [% v/v]	>25
pH value	6.5–8.5
SO_4^{2-} ppm [d.w.]	>500
NH_4^+ ppm [d.w.]	<400
NO_2^- ppm [d.w.]	<0.1
NO_3^- ppm [d.w.]	No limited
P total [% d.w.]	>0.3
N total (Kjeldahl nitrogen) [% d.w.]	>0.5
Organic matter content [% d.w.]	>15
TOC [% d.w.]	>7
Respiration activity within 7 days [mg O_2 g^{-1} o.d.w.]	<8

d.w.—dry weight.
o.d.w.—organic dry weight.

it is recommended to mix the compost produced from fine structure material, such as sewage sludge, with coarse material that have an impact on the formation of structures, such as woodchips or bark. This treatment provides a suitable air porosity that should range from 30 to 45% vol., and moisture content of 40–50% by weight (Huber-Humer et al. 2008). Moreover, inert materials such as perlite, keramzite and glass beads or polystyrene pellets may be used as the material that influences the formation of structures (Melse & Werf, 2005, Abichou et al. 2006a, Powelson et al. 2006, De Visscher et al. 1999, Pawłowska et al. 2011). They have a rigid structure, are not biodegradable; which increase the resistance of the bed to sedimentation, reduce susceptibility to changes in the porosity over the course of time, and consequently prolong the life of the filter bed material.

The material used as carrier of microorganisms must contain appropriate nutrients concentration, mainly nitrogen and phosphorus. Huber-Humer et al. (2009) recommend that the content of Kjeldahl nitrogen in the compost used to form landfill biocovers should not be lower than 0.5% d.w; whereas, P_{total} should not decrease below 0.3% d.w. The literature data indicate that also the excess of certain forms of biogenic elements, such as nitrogen in the ammonium form, may have a negative influence on the rate of oxidation of these compounds; whose decomposition is catalysed by methane monooxygenase. NH_4^+ ions compete for the MMO active sites and by blocking them they decrease the efficiency of oxidation process of methane (Albanne et al. 2007) and other compounds that can be co-metabolised by methanotrophs. Another important element, whose content in the substrate used in the biofiltration of landfill gas should be controlled, is copper. When the concentration of Cu^{2+} exceeds the value of 1 μM, the inhibition of sMMO formation occurs (Hanson & Hanson, 1996). sMMO is present in cells of the type II methanotrophs, which are characterized by high methanotrophic activity and low affinity with CH_4. Therefore, while choosing the material for landfill covers or methanotrophic biofilters, it should be noted that the copper concentration in the soil solution did not exceed the value of 1 μM.

5.5.2 *Temperature of microorganisms growth*

The temperature of the filter bed material for microbial growth is a resultant value of multiple components. Not only does it depend on the temperature of the environment, but also on the thermal properties of the filter bed material itself, that is heat capacity or heat transfer coefficient. Apart from heat diffusion from the atmosphere, an energy released in the reactions occurring in the filter material (e.g., by the oxidation of 1 mole of methane, there are 880 kJ of energy emitted (Nikiema et al. 2007)), energy from organic matter mineralization processes, heat from landfill gas, and sometimes heat from heat exchangers, might all be the heat source in the material. A temperature gradient in the bed profile is also influenced by the gas flow rate through the bed, and the material of the bed (Humer & Lechner, 1999).

Most microbes growing in landfills covers prefer mesophilic conditions, and the optimum temperature ranges from 25 to 35°C. Numerous microorganisms growing in different materials exposed to high concentrations of methane were found to be mesophiles (Whalen et al. 1990, Boeckx & Van Cleemput, 1996, Visvanathan et al. 1999, Streese et al. 2001, Mor et al. 2006). The type I methanotrophs tolerate lower temperatures better than the type II methanotrophs (Börjesson et al. 2004), which directs particular attention on the temperature control in the deposits at the removal of methane. The type II methanotrophs show low affinity with methane and high oxidation activity. Moreover, the excessive temperature rise is not beneficial for methanotrophs inhabiting landfill covers. At temperatures above 40°C, the activity of methanotrophic bacteria isolated from the landfill cover decreased, and at 50°C it was completely disappearing (Zeiss, 2006). The effect of the temperature on methanotrophic activity is significant. Vant't Hoff temperature coefficients Q_{10}, which specify the growth of rate of the reaction at a temperature change of 10°C, ranged from 1.9 to 8.4 (Whalen et al. 1990, Zeiss, 2006, Einola et al. 2007), depending on the soil type, its properties and range of temperatures tested.

5.5.3   *Composition of gas mixture*

An important factor influencing the efficiency of the biological oxidation methods in relation to the LGF is the complex chemical composition of the gas mixture, which is site- and time-dependent. An analysis of the relationship between CH_4, NMOC, and individual organic compound emissions from landfill suggests a correlation between CH_4 and trace organic oxidation (Barlaz et al. 2004). Many organic compounds present in the biogas can be used as a source of carbon and energy by the same species of microorganisms by competing with each other for the active site of enzymes. Some compounds may inhibit the decomposition of other compounds. They may act as cosubstrates for specific groups of bacteria; which, even though are subjected to chemical changes due to the action of enzymes, they are not assimilated by them. Only the products of these transformations, being less specific, are a source of carbon and energy for other organisms. Moreover, certain compounds have a toxic influence on specific microorganisms limiting their participation in the process.

The search for the interaction between the mutual influence of compounds undergoing biofiltration is very important from the point of view of the efficiency of the use of biological methods for the Volatile Organic Compounds (VOCs) removal from biogas. The results of studies regarding the ability of removal of trace gases in the presence of methane, which is an essential component of LFG, are extremely important. Research indicates that methane can accelerate the decomposition of other hydrocarbons present in LFG, such as benzene and toluene. Lee et al. (2010) stated that the rate of oxidation of benzene increases two-fold in the landfill cover soil, and *ca.* 2.3-fold in the wetlands soil, when compared to the rates of oxidation of these compounds during soils incubation in the gas mixture without methane. The influence of methane on toluene oxidation was less significant in landfill soil (1.4-fold increase) but just as significant in wetland soil (2.4-fold increase). The increase in the rate of trace compounds removal from the gas mixture implies the reduction in the rate of methane removal (Lee et al. 2010), which is associated with the competition for the active sites on the MMO.

5.6   TECHNOLOGICAL APPROACH TO APPLICATION OF BIOLOGICAL METHODS FOR MITIGATION OF LFG EMISSION

Diversity of microorganism and their metabolic pathways occurring in the cells provides many opportunities to use biooxidation processes for mitigation of LFG emission on different types of MSW landfills. The choice of a specific technological solution depends on the conditions prevailing in the landfill. Moreover, biofiltration can be used as a main or supplementary method in relation to the thermal utilization of gas. What is more, it can be used in both landfills with active and passive gas collection systems, mainly in post closure phase. However, its use during active phase is not excluded. It is also possible to use biofilters in old, unsecured or secured ineffectively landfills; as well as in landfills, where waste from mechanical-biological pre-treatment is stored. The choice of technology and the parameters of biofilter should depend on the type of landfill: sanitary landfills equipped with bottom liners and capping system, sanitary landfills under operation, open dumps; expected LFG production and its composition; as well as legal requirements.

5.6.1   *Forms of biotic systems for landfill gas mitigation*

5.6.1.1   *Landfill biocovers*
Biocover is the easiest and the cheapest biological method of LFG emission mitigation. It is a biologically active layer of porous material covering the landfill, through which the landfill gas passes. Biocover is usually placed on a drainage layer, for instance made of gravel (Fig. 5.8a), which allows for a more equal distribution of the gas in the bed. The treatment of gas does not require any particular preparation of the object. Low surface loading rate of pollutants,

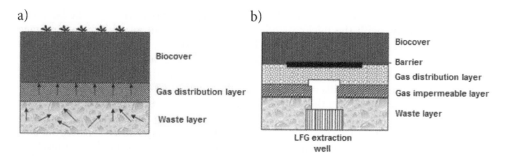

Figure 5.8   Landfill biocovers: a) overlaid the drainage layer (Huber-Humer et al. 2008), and b) integrated with landfill degassing system (Ettala & Väisänen, 2000).

resulting from the distribution of gas emissions on a large surface of the filter bed material, creates the possibility of the use of materials with lower methanotrophic activity, e.g. soils occurring in the surrounding area. Moreover, biocover serves as a recultivation layer.

However, the use of biocover as the final covers on sanitary landfills may bring some limitations. Although the Directive 1999/31/EC does not insist on the application of tight capping in the case of landfills for non-hazardous waste, together with the ones for municipal solid waste; it requires landfill gas recovery. For this purpose, landfill surface sealing, which increases the effectiveness of recovery and improves the quality of landfill gas, is necessity. A soil layer, applied over the surface sealing, aiming to be recultivation layer on the landfill cannot function as a biocover, because the landfill gas does not have any access to it. Recently, the biocover, used as final cover, finds its application on the old landfills or the ones that are improperly secured, which gas with a low methane content and in small amount are emitted from. It can also be used in landfills producing small amounts of LFG. What is more, the biocover can be a complementary solution to the degassing system (Huber-Humer et al. 2008). The disadvantage of this solution is the high consumption of material constituting the filter bed and the lack of control of process conditions.

Specific version of biocover, which might be used in sanitary landfills, is biocover integrated with degassing system of the landfill (Fig. 5.8b). This technology has been patented (WO/2001/017701) in Finland (Ettala & Väisänen, 2000; Ettala & Väisänen, 2002). It involves the use of biocovers in landfills covered by a layer that is impermeable for gases which is also equipped with gas extraction well. The distribution of gas takes place by perforated tubes arranged above the insulation layer, in a biological filter bed. Gas from the gas extraction wells flows from the drainage layer along the barrier which is placed perpendicular over the well, on the border of the drainage layer and soil biocover. It reaches the highly porous material after passing the barrier. Such distribution of the gas stream from the well causes the dispersed flow of the gas through the drainage layer; and then, the gas lighter than air flows upward to the biocover, where it is oxidized. Such a solution were tested in a technical scale in the landfill in the southern part of Finland, which operated in the years 1987–2001. Biocover integrated with tight landfill cover removed more than 46% of the methane emitted from the landfill during the summer, and at least 25% in the winter (Einola et al. 2009).

This type of biocover can provide a self-sufficient (in the absence of gas collection for energy purposes), or complementary solutions to thermal utilization (after cessation of gas combustion).

According Huber-Humer et al. (2008), in climate conditions in the Central Europe, the most effective, in both terms of the efficiency of the methane oxidation and the prevention of landfill leachate, is the landfill cover consisting of two layers: the top layer with the thickness of 1.2 m, made of the mature compost characterized by a low value of respirometric activity ($BOD_7 < 8$ mg O_2 g^{-1} d.w); and the lower layer with a thickness of 0.30–0.5 m, made of the coarse gravel, which is established to unify the flow of gas to the upper layer.

Such a cover allows for the removal of 95–99% of CH_4 that passes through the bed. The compost layer alone, with the thickness of 0.30–0.40 m, without an underlying drainage layer allows for the removal of 68–74% of CH_4. In the two-layer cover, no decrease in the capacity of CH_4 oxidation in the winter season was observed, as opposed to the case of traditional monolayer cover (Börjesson et al. 2001). In Austria, there are at least 5 landfills or separated sections of landfills covered by two-layer cover. Six-year monitoring of these landfills leads to the conclusion that their effectiveness equals almost 100% (Huber-Humer et al. 2008).

Research conducted under subtropical climate (Florida, USA) also confirmed the efficiency of a two-layer cover, which includes a layer of mature compost made of household and green waste with a thickness of 0.50 m; and a drainage layer of crushed glass with a thickness of 0.10–0.15 m. This landfill cover was placed on a low effective (in terms of methane removal) cover, which constitutes the layer of loamy sand and sandy loam, of a thickness of 0.65–0.75 m (Bogner et al. 2005, Abichou et al. 2006b). By increasing the thickness of the cover, retention time of gas in the layer is extended, which together with improved humidity allowed for a 10-fold decrease in methane emissions from landfills (Stern et al. 2007). The use of plants prevents the transpiration from the surface of the cover; whereas, the root systems support the transport of atmospheric oxygen to the soil, and improve aerobic conditions in the cover. Metabolic products released to the root zone are the source of nutrients for methanotrophic bacteria stimulating their development (Tanthachoon et al. 2007).

In special cases, landfill biocover may also remove methane from the atmospheric air. Such a phenomenon occurs when negative pressure generated by the active degassing system leads to the situation when the atmospheric air is sucked inside to the landfill through the layers of the cover (Bogner et al. 2005).

According to the provisions concerning the operation of landfills, a user is obliged to use intermediate covers, and sometimes daily covers, which results from the need to limit the influence of the landfill in an active phase on the atmosphere. In a case when only a part of landfill is exploited, the rest of the landfill may be temporarily out of use and secured due to the greenhouse gas. Usually, 0.15 m of a layer of locally available or contaminated soils, compost, sewage sludge and artificial cover (Haughey, 2001, Huber-Humer, 2008) or construction waste are applied for this purpose. These covers protect the environment primarily against the waste and dust migration, animals access, and to some extent they also reduce the greenhouse gas emissions. The temporary covers that are applied in order to effective reduce of methane emissions are called thin biocovers. For instance, layers with a thickness of 0.30 m formed from compost mixed with a small amount of sawdust (up to 10% by weight) may play such a role (Perdikea et al. 2008).

5.6.1.2 *Biowindows*

Biowindows may be applied on sanitary landfills, tightly covered or landfills that were earlier covered with material with low efficiency in terms of the oxidation of organic compounds emitted from the landfill (Fig. 5.9a). Biowindows are separated pieces of landfill capping system filled with a bioactive material wit high porosity. Landfill gas encountering the layer with low resistance, relative to the remaining part of the cover, migrates to a biowindow, where some components are oxidised. Such solutions are widely practiced in connection with bentonitic covers. Compost is usually the material filling biowindows (Huber-Humer et al. 2008). Biowindows can also be applied in landfills with low emissions of gas; for instance, when the active gas collection system is turned off. The advantage of this solution is the fact that the consumption of the filter bed material is small, in comparison with the biocover. However, due to the high surface loading rate, the material must be highly active taking the oxidation of pollutants into consideration. The other disadvantage of biowindows is the lack of control of process conditions.

5.6.1.3   *Biofilters*

Biofilters are separated chambers filled with material that provides a high removal efficiency of pollutants. Compost, peat, soils or mixtures of these materials with perlite, expanded clay or polystyrene pellets may constitute the filter material. They are located in the place of the controlled discharge of gas, in the end parts of degassing system, or are built-up into the land-fill cover (Fig. 5.9b). Biofilters find very broad applications. They can operate at both passive (when gas flows under its own pressure), and active degassing system of landfill (when gas is sucked in). The use of biofilters is possible in the case of new landfills equipped with the bottom liners, and where there is no rational justification for the biogas recovery for energy purpose under such circumstances (Huber-Humer et al. 2008). Such conditions may occur after the reduction of gas production, which proceed with the age of the landfill; or on small landfills, where the time of gas generation with a high concentration of methane is too short to allow for a profitable installation of degassing system (Pawłowska, 2008). Biofilters can be used as a complementary method for biogas recovery system; for instance, in the case of a drop in gas quality that would justify the cost-effectiveness of its combustion or in the case of energy-generating equipment failure or breakdown of flare. The use of biofilters is also pos-sible in the course of operation of the landfill; for example, in cells that are temporarily out of operation (Huber-Humer et al. 2008). One of advantages of biofilters application is a small consumption of filler bed material, which entails greater possibilities to choose the material. However, the high surface loading rate of biofilter compared with the loading of biocovers involves the selection of the filter material providing a high methane-oxidizing capability. Furthermore, there is the possibility of exchanging the bed after lowering its activity.

The biofilter can be classified as an open (Fig. 5.9b) and closed (Fig. 5.10) systems. In open biofilters which are mostly found in landfill sites, gas stream flows upwards, while the O_2 diffuses from the ambient air into the bed (Nikiema et al. 2007). However, the aeration of filtered gas is also possible (Streese & Stegmann, 2003). The main disadvantage of this solution is a difficulty to maintain steady operational parameters, due to the lack of the con-trol of temperature, moisture content, and air accessibility. Closed biofilters do not possess these drawbacks. In their case both the purified gas and the air stream are supplied forcibly; whereas, the humidity and temperature can be suitably controlled. Humidity control is done by the gas humidification (Du Plessis et al. 2003, Nikiema et al. 2005) or sprinkling of the bed (Powelson et al. 2006, Tanthachoon et al. 2007). Oxygen can be supplied with a stream of biogas (Streese & Stegmann, 2003) or introduced directly into the bed, on its different depths (Haubrichs & Widmann, 2006). The full process control and independence from external conditions is a big advantage of closed biofilters; however, high operating costs associated with the need to supply energy to the system are their serious disadvantage. Open systems are usually at least 15% cheaper than closed systems (Huang et al. 2011).

There are some possibilities to reduce the impact of external conditions, such as too high air temperature, on biofilter operation. For instance, materials with different granulometric

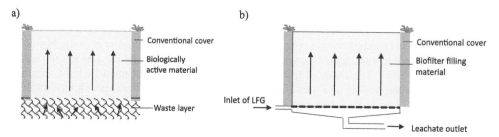

Figure 5.9   Landfill gas treatment system: a) biowindow, and b) open biofilter integrated with the landfill cover (Huber-Humer et al. 2008, modified).

a)

b)

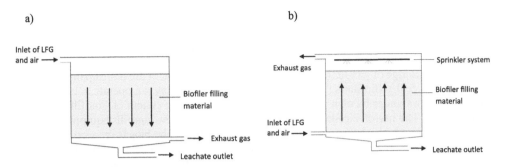

Figure 5.10   Closed biofilters: a) with gas supply from the top (Huber-Humer et al. 2008; modified); and b) with the gas supply from the bottom (Bilitewski et al. 2003, modified).

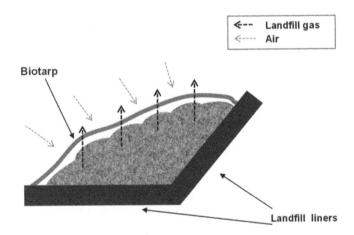

Figure 5.11   Schematic arrangement of biotarp on the layer of waste (Huber-Humer et al. 2008).

composition can be used to prevent rapid drying of an open biofilter. Placement of the fine grained material with a large water retention near the surface forms a barrier against drying of the coarse material situated below, which is gas-permeable (Powelson et al. 2006). The use of filter materials with a high content of organic matter is also the agent hedging them for drying. The surface layer of the organic material becomes hydrophobic after drying. It leads to the decrease of its capillary capacities, preventing transpiration from the surface (Huber-Humer et al. 2008). What is more, the vegetation that could be placed on the surface of an open biofilter prevents the water evaporation.

5.6.1.4   *Biotarps*

An innovative way to apply biotic systems to mitigation of LFG emission is to use biotarps. Biotarp is a flexible, thin layer biologically active film, which can be used as a daily cover of waste (Fig. 5.11). The material from which biotarp is produced must be light, flexible, resistant to repeated rolling and unrolling, retaining moisture, have high specific surface area and porosity (Hilger et al. 2007), and be resistant to biodegradation. For technical reasons, biotarp cannot be too thick and, at the same time, must provide a suitable methane retention time in the bioactive layer (Huber-Humer et al. 2008). Among the tested materials; such as: natural sponge, geotextile, polycarbonate membrane, non-woven polypropylene geotextile, glass beads; the greatest suitability for use in creating biotarps possess a natural sponge and geotextile (Hilger et al. 2007). They had the greatest ability to remove methane and absorb methanotrophic microorganisms. What is more, methanotrophs were immobilized well on calcium alginate gel; however, this bed tend to dry quickly. Selection of bacterial strains used for inoculation of mat is extremely important. Not all methanotrophs are able to survive 12 hour interruption of the access to CH_4, which is significant because the mats are rolled for a long part of the day (Huber-Humer et al. 2008).

5.7 OPERATING AND CONTROL PARAMETERS OF LANDFILL GAS BIOFILTERS

Controlling the amount of gas supply, air and/or water supply, maintaining a stable temperature or pH of the bed; these are the conditions that allow to maximise the potential of microorganisms inhabiting the filter beds. Those conditions provide optimal growth for bacteria that degrade pollutants. However, not all technical solutions for provide the ability to control the conditions of the process. In this respect the most advanced are active biofilters. Apart from the selection of the bed material, it is also possible to control the amount of gas and air supplied.

The main operating parameters of biofilters performance are as follows:

Surface loading rate [SLR]—indicates the volume of the treated gas that is let through the unit of surface area of biofilter per unit of time.

$$SLR = \frac{Q}{A} \quad [m^3\ m^{-2}\ h^{-1}]$$

where Q = gas flow rate [$m^3 h^{-1}$]; A = total surface of filter material in the biofilter [m^2]. Volumetric loading rate [VLR]—indicates the volume of the treated gas that is let through the biofilter per unit of its volume and unit of time.

$$VLR = \frac{Q}{A} \quad [m^3\ m^{-3}\ h^{-1}]$$

where V = volume of the filter material in the biofilter [m^3]
Mass loading rate [MLR]—indicates the amount of the pollutant which is introduced into the biofilter per unit of its volume or surface area, and per unit of time.

$$MLR = \frac{Q \times C_{in}}{V} \quad [g\ m^{-3}\ h^{-1}]$$

$$MLR = \frac{Q \times C_{in}}{A} \quad [g\ m^{-2}\ h^{-1}]$$

where C_{in} = concentration of the pollutant in the inlet gas stream [g m^3].
Empty bed residence time (EBRT)—gas residence time in the biofilter, calculated in relation to the total volume of the bed, according to the equation:

$$EBRT = \frac{V \times 3600}{Q} \quad [s]$$

Typical EBRT for common biofilter applications is from 25 s to over a minute (Devinny et al. 1999, Melse & Van De Werf, 2005). For example, recommended EBRT for gas emitted from composting facilities is from 30 to 60 s (Haug, 1993). But biofilters for treatment of landfill gas which contains methane in the concentrations up to 260 g/m^3 (40%, v/v) have been operated at high EBRT ranging between 5 min and 5 hours (Melse & Van De Werf,

2005). This is justified by the relatively high methane concentrations in the gas emitted from landfill sites, and low methane solubility in water (Bunsen solubility coefficient $\alpha_{20} = 0.0331$ dm^3dm^{-3}—Gliński & Stępniewski, 1985).

True (or actual) residence time (τ)—gas residence time in the biofilter, calculated in relation to the volume occupied by void space within filter bed material

$$\tau = \frac{V \times \varepsilon \times 3600}{Q} \quad [s]$$

where ε = porosity of the packing material [-].

Removal efficiency and elimination capacity of particular pollutants from treated gas stream are the parameters used to assess the effectiveness of biofiltration process.

Removal efficiency (RE)—is the ratio of the amount of pollutant removed in the biofilter to the amount of pollutant fed into biofilter, and it is calculated as follows:

$$RE = \frac{C_{in} \times C_{out}}{C_{in}} \times 100 \quad [\%]$$

where RE = removal efficiency [%]; C_{in} = the concentration of removed gas in the gas mixture entering the biofilter [g m^{-3}]; C_{out} = the concentration of removed gas in the gas mixture leaving the biofilter [g m^{-3}].

Elimination capacity (EC)—gives the amount of pollutants removed in volume unit of biofilter per unit of time:

$$EC = \frac{Q \times (C_{in} - C_{out})}{V} \quad [g\ m^{-3}\ h^{-1}]$$

It should be also calculated as:

$$EC = MLR \times \frac{RE}{100} \quad [g\ m^{-3}\ h^{-1}]$$

In the literature concerning the biological removal of methane in systems simulating landfill conditions, the mass of substance degraded per surface unit of filter material per time unit (e.g. g m^{-2}h^{-1}) is the most frequently used parameter. It is called methanotrophic/methane oxidation capacity, methane oxidation/methane uptake rate or methane oxidation potential. The relation of the pollutant removal to the surface of the bed instead of its volume is pragmatic. In order to ensure the effective methane removal from the landfill, it is necessary to provide a large amount of filter material; whereas, the height of beds should be established in advance. It should not exceed 1 m due to the diffusive nature of the oxygen supply. Therefore, the assessment of the methanotrophic capacity of the bed in relation to its surface rather than volume is more reliable.

5.8 QUANTITATIVE APPROACH TO METHANE AND VOCs REMOVAL IN LANDFILL COVERS AND BIOFILTERS

The results of the studies on methane oxidation capacity presented in Table 5.7 show, in this regard, very large disproportions in materials. Until now, many kinds of materials were tested: organic (different types of composts, peat); mineral (soils, sand and gravel materials); and organic-mineral mixtures (e.g. compost with expanded clay pellets, perlite, or glass cullet). Most of the studies were carried out under laboratory conditions, but some of them were also performed in field scale (Gebert & Gröngröft, 2006b, Powelson et al. 2006, Dever et al. 2011).

Table 5.7 Methanotrophic capacities of different materials exposed to high methane concentration.

Material type	Mass loading rate of methane	Methane elimination capacity	Details	References
Coarse sand from landfill cover	274 g m^{-3}d^{-1}	155.9 ± 1.5 g m^2 d^{-1}	Laboratory experiment. Passive aeration. Steady state values	Kightley et al. (1995)
Fine sand from landfill cover		102.4 ± 4.5 g m^2 d^{-1}		
Loamy clay from landfill cover		101.7 ± 4.5 g m^2 d^{-1}		
Arable soil	214 g m^{2}d^{-1}	170 g m^2 d^{-2}	Laboratory experiment. Passive aeration. Averages values	De Visscher et al. (1999)
Soil from landfill cover	368 g m^{2}d^{-1}	240 g m^2 d^{-1}		
Agricultural loam	~320 g m^{-2}d^{-1}	150/107 g m^2 d^{-1}	Laboratory experiment. Passive aeration. Maximum values/Steady state values	Stein & Hettairatchi (2001)
Landfill loam		301/180.6 g m^2 d^{-1}		
Sedge peat		160.6/96.3 g m^3 d^{-1}		
Coarse sand	300 g m^{2}d^{-1}	152.1 ± 7.1 g m^2 d^{-1}	Laboratory experiment. Passive aeration. Averages values.	Pawłowska et al. (2003)
Fine gravel		135.4 ± 8.2 g m^2 d^{-1}		
Leave compost	~500 g m^{-2}d^{-1}	400/120 g m^{-2}d^{-1}	Laboratory experiment. Passive aeration. Maximum values/Steady state values	Wilshusen et al. (2004b)
Garden waste compost		50/40 g m^{-2}d^{-1}		
Composted wood chips		250/120 g m^{-2}d^{-1}		
MSW compost		250/100 g m^{-2}d^{-1}		
Green waste compost	288–3120 g m^{3}d^{-1}	1500/214 g m^3 d^{-1}	Laboratory experiment. Active aeration. Maximum values/Averages values	Streese & Stegmann (2003)
Mixture of compost, bark and peat (equal volumes)		960/280 g m^3 d^{-1}		
Multilayer biofitler (compost and bark)		720/234 g m^3 d^{-1}		
Expanded clay pellets overlain by soil	0–6000 g m^{3}d^{-1}	1900 g m^3 d^{-1}	Field experiment. Open biofilter. Passive aeration	Gebert & Gröngröft (2006b)
Inorganic packing material	1700 g m^{-2}d^{-1}	720 g m^3 d^{-1}	Laboratory experiment. Active aeration. Maximum values	Niekiema et al. (2005)
Mature compost		360 g m^{3}d^{-1}		
Green waste compost with polistyrene pellets (1:1 v/v)	250–500 g m^{-2}d^{-1}	242 g m^{-2}d^{-1}	Field experiment carried out on model landfill gas. Passive aeration. Averages values	Powelson et al. (2006)
Coarse sand overlain by fine sand (water spreading)		203 g m^{-2}d^{-1}		
Yard waste compost	691 g m^{-3}d^{-1}	663 g m^{-3}d^{-1}	Laboratory experiment. Active aeration. Average value	Haubrichs & Widmann, (2006)
Composted pine bark	<420 g m^{-2}d^{-1}	RE > 70%	Laboratory experiment. Active aeration	Du Plessis et al. (2003)
MSW compost	1044 g m^{-2}d^{-1}	1036 g m^{-2}d^{-1}	Laboratory experiment. Active aeration. Maximum values	Pawłowska et al. (2011)
Horticultural substrate		889 g m^{-2}d^{-1}		

RE: removal efficiency.

The maximum values of methane oxidation capacity measured in the material (60–80 cm of depth) under laboratory and field-scale tests, in which the oxygen was diffused to the filter bed from the atmosphere (passively aerated biofilter) ranged from 200 to 400 g m^{-2}d^{-1} (De Visscher et al. 1999, Stein & Hettiariatchi, 2001, Wilshusen et al. 2004b, Powelson et al. 2006, Pawłowska & Stepniewski, 2011). But, when the oxygen was actively supplied to the biofilter (actively aerated biofilter) the methanotrophic capacity increased up to a value of 1000 g m^{-3}d^{-1} (Streese & Stegmann, 2003, Pawłowska et al. 2010). This is because the most of methane-oxidizing bacteria are obligatory aerophiles (Mancinelli, 1995). Therefore, the availability of oxygen is a key parameter that differentiates the efficiency of methanotrophic biofilters. Methanotrophic potential measured in biofilters, in which the access of oxygen to the microorganisms was not limited (in actively aerated biofilters), depended on a limited scale on the type of the bed material. Despite strong differences in the parameters of the filter beds, such as organic matter content (ranged from 9.5 a 85.5% dry weight) and water holding capacity (ranged from 122 is 349% dry weight), the ability to oxidize methane varied within a narrow range from 839 to 1022 g m^{-2}d^{-1}. The difference between minimum and maximum value was about 20%. In contrast, the maximum capacities of the materials filling up the bio-filters aerated *via* diffusion in wider interval ranged from 164 is 399 g m^{-2}d^{-1}. In this case the difference between minimum and maximum value was about of about 140% (Fig. 5.12). The most important parameters influencing the process in these biofilters were: porosity, water holding capacity and pH. Increase of the total porosity (in the range 60.6 to 78.7% v/v) and pH value (from 6.0 to 6.8) enhanced the methanotrophic activity, while the increase of water holding capacity (from 97.5 to 168% d.w) decreased the methane oxidation probably due to deterioration of gas diffusion conditions. Only the influence of C:N ratio was less significant in passively aerated biofilters than in actively aerated one (Pawłowska & Stępniewski, 2011).

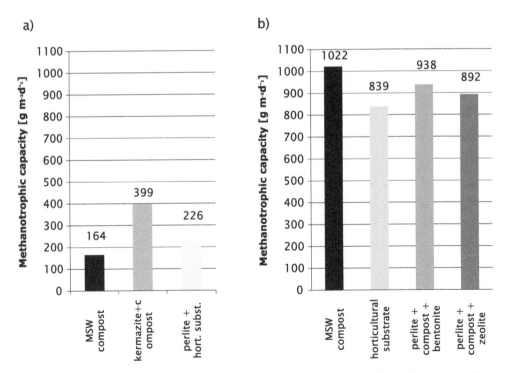

Figure 5.12   Maximum methanotrophic capacities measured in continuously flushed by methane laboratory biofilters, a) actively aerated; b) aerated by oxygen diffusion from atmosphere (Pawłowska & Stępniewski, 2011).

In the case of biofilters without forced aeration (passively aerated), organic matter content in the soil used in the biofiltration of landfill gas plays an important role. Materials rich in organic matter, such as compost or a mixture of organic and mineral fertilizers, are characterized by the methane oxidation capacity that ranges from 160 to 400 g CH_4 $m^{-2}d^{-1}$ (Powelson et al. 2006, Wilshusen et al. 2004b, Pawłowska & Stepniewski, 2011); whereas, the methane oxidation capacity in mineral soils, sands and gravels ranges from 102 to 240 g CH_4 $m^{-2}d^{-1}$ (Kightley et al. 1995, De Visscher et al. 1999, Pawłowska et al. 2003). Moreover, peats are organic materials used in the biofiltration of gases. They exhibit lower values of CH_4 oxidation than composts. Maximum methanotrophic capacity of sedge peat was about 240 dm^3 $m^{-2}d^{-1}$ (Stein & Hettiaratchi, 2001) and was similar to the one found in arable soil, which was studied by these authors.

Taking the capabilities of removing methane in soil materials into consideration, it is estimated that in order to ensure the oxidation of methane at flow rates in the range of 0.01–2.5 m^3h^{-1} the size of the biofilter should be at least 1 m^3 of filter bed volume (Straka et al. 1999, Stresse & Stegmann 2003, Haubrichs & Widmann, 2006). The height of the open biofilters with passive ventilation, used for CH_4 elimination, must be lower than 1 m (Boeckx & Van Cleemput, 2000, Stein & Hettiaratchi, 2001, Park et al. 2002).

There are little data concerning the capacity of trace gases removal from landfill gas *via* biofiltration. The laboratory study carried out in dynamic system simulating landfill cover permeated with artificial gas (CH_4:CO_2 = 1:1 v/v) containing VOCs showed the highest elimination capacity, equal to 0.73 g $m^{-2}d^{-1}$ for dichloromethane. ECs of benzene and toluene were 0.18 and 0.12 g $m^{-2}d^{-1}$, respectively (Scheutz & Kjeldsen, 2005). Vinyl chloride was degraded with the capacity of 0.18 g $m^{-2}d^{-1}$. When it comes to tetrachloromethane, trichloromethane and trichloroethylene; they were degraded with low capacities ranged from 0.023 to 0.041 g $m^{-2}d^{-1}$ (Scheutz & Kjeldsen, 2005). The ability of landfill cover soils to biodegrade many trace gases was observed also under field conditions (Scheutz et al. 2008, Bogner et al. 2010). Substantial reductions for aromatics, alkanes, and lower chlorinated compounds were noticed by Bogner et al. (2010). The non-methane hydrocarbon emissions from the surface of landfill biocover consisted of 60 cm ground garden waste placed over a 15 cm gas distribution layer (crushed glass from discarded fluorescent lights) was in the range of 10^{-9} to 10^{-3} g $m^{-2}d^{-1}$.

The EC values measured by Scheutz & Kjeldsen (2005) were very low as compared to the elimination capacities of trace gases measured in laboratory biofilters purged with contaminated air (free of methane, and rich with oxygen), and inoculated by specific VOCs degrading microorganisms. For example maximum toluene elimination capacity from a synthetic waste air stream in bench-scale biofilter (0.5-m high filter bed), inoculated with a toluene-degrading strain of *Acinetobacter* sp., was 242 g $m^{-3}h^{-1}$ (Zilli et al. 2000). Elimination capacity for toluene in biofilter filled with cubic polyurethane foam inoculated with *Stenotrophomonas maltophilia* was approximately 350 g $m^{-3}h^{-1}$; and it was 2 to 3.5 times higher than for the biofilters filled with porous ceramic, lava, and granular activated carbon mixed with ceramic (Ryu et al. 2010). Maximum toluene elimination capacity in laboratory-scale up flow biofilter, packed with compost-based filter material was 100 g $m^{-3}h^{-1}$ (Delhomenie et al. 2001).

Such low EC values in the case of trace gases removal from landfill gas may be linked to oxygen limitation and the presence of methane. Methane enables growth of methanotrophs, which leads to the degradation of trace gases on cometabolic pathways.

Batch tests showed that methane significantly influences the rate of VOCs removal (Table 5.8). Lee et al. (2010) revealed that methane stimulated benzene and toluene oxidation in landfill cover soils. The rate of benzene removal increased from 0.92 μmol g^{-1} dry soil h^{-1}, when soil was incubated in benzene/toluene mixture to 1.82 μmol g^{-1} dry soil h^{-1}, when soil was incubated in the gas mixture enriched with 5% of methane. The rate of toluene degradation increased from 0.91 mol g^{-1} dry soil h^{-1} to 1.36 μmol g^{-1} dry soil h^{-1}, respectively. They noticed that methane oxidation rates decreased with increasing benzene/methane ratio. When the methane was the only carbon source the rate of its oxidation was 5.56 μmol g^{-1} dry soil h^{-1}; however, when the toluene and benzene was added, the rate decreased to 3.42 μmol

Table 5.8   Rates of CH$_4$ and selected NMOCs removal by microorganism inhabiting landfill covers.

Material/experimental conditions	Compounds							References
	Methane (M) μg g^{-1}d^{-1}	Benzene (B) μg g^{-1}d^{-1}	Toluene (T) μg g^{-1}d^{-1}	Xylene μg g^{-1}d^{-1}	1,1,1-Trichloroethane TCA μg g^{-1}d^{-1}	Trichloroethylene (TCE) μg g^{-1}d^{-1}	Vinyl chloride (VC) μg g^{-1}d^{-1}	
Landfill cover soil/temp. 25°C	50–250	2.5–8	1.9–8	–	–	–	–	Kjeldsen et al. (1997)
Landfill cover soil/temp. 10°C	–	–	–	–	<0.01–0.03	0.03	–	
Landfill cover soil/temp. 22°C incubated with M and selected trace compounds	18-34	27.9	38.7	–	–	0.057	8.56	Scheutz et al. (2003)
Landfill cover soil/temp. 22°C incubated with M and selected trace compounds	576–2688	20.2	33.4	4.1	no degradable	1.4	34.9	Scheutz et al. (2004)
Landfill cover soil incubated with mixture of B & T	–	1725* (0.92 μmol g^{-1}$_{d.w}$ h^{-1})	2012* (0.91 μmol g^{-1}$_{d.w}$ h^{-1})	–	–	–	–	Lee et al. (2010)
Landfill cover soil incubated with mixture of B, T & M	1317* (3.42 μmol g^{-1}$_{d.w}$ h^{-1})	3412* (1.82 μmol g^{-1}$_{d.w}$ h^{-1})	3008* (1.36 μmol g^{-1}$_{d.w}$ h^{-1})	–	–	–	–	
Landfill cover soil incubated with M	2140.4* (5.56 μmol g^{-1}$_{d.w}$ h^{-1})	–	–	–	–	–	–	
Consortium of bacteria from landfill cover-incubated with single substrate	754.5 × 10³** (1.96 mmol g^{-1}$_{DCW}$ h^{-1})	281.1 × 10³** (0.15 mmol g^{-1}$_{DCW}$ h^{-1})	1702.8 × 10³** (0.77 mmol g^{-1}$_{DCW}$ h^{-1})	–	–	–	–	Lee et al. (2013)

* related to dry weight of soil.
** related to dry cell weight (DCW).
Original units are given in parenthesis.

g^{-1} dry soil h^{-1} (Lee et al. 2010). Drop of methane oxidation rate due to a presence of such aromatics as m-xylene, p-xylene and ethyl benzene was observed also in pure culture *Methylocystis* sp. (Lee et al. 2011b).

Contrary to the results obtained by Lee et al. (2010), methane had an insignificant effect on benzene or toluene degradation in the study carried out by Lee et al. (2013). In this experiment landfill cover soil was used as an inoculum source to enrich a methane, benzene, and toluene-degrading consortium MBT14. The consortium was able to simultaneously degrade methane, benzene and toluene (Table 5.5). However, the coexistence of benzene and toluene inhibited the methane degradation rates.

According to Scheutz et al. (2004) lower chlorinated compounds are biodegradable in the presence of methane and oxygen. Degradation rates of the chlorinated aliphatics are inversely related to the ratio of chlorine and carbon. Vinyl chloride was removed with higher rate than trichloroethylene (TCE). Fully halogenated hydrocarbons (PCE), tetrachloromethane (TeCM), chlorofluorocarbon (CFC-11, CFC-12, and CFC-113) were not biodegraded (Table 5.5). Maximal oxidation rates for the halogenated aliphatic compounds varied from 0.7 to 41 μg $g^{-1}d^{-1}$, and for aromatic hydrocarbons from 4.1 to 38.7 μg $g^{-1}d^{-1}$ (Scheutz et al. 2003, Scheutz et al. 2004).

5.9   CRITICAL APPROACH TO LFG BIOFILTRATION

The use of biological methods in mitigation of landfill gas emission is connected with certain problems, the intensity of which depends among others on the technology used. The most important obstacle is time-dependent decrease in oxidising-activity of filter bed, associated with insufficiency of oxygen availability for microorganisms. That is caused by the deterioration of air diffusion inside the filter bed due to decline in total and gas porosities of the filter bed material. The formation of Extracellular Polymeric Substances (EPS), which are the metabolic products of bacteria inhabited the filter bed, is one of the reasons of the decline in porosity. The accumulation of these substances causes clogging of the pores and hinders the supply of gases to the bacteria cells. These substances are composed of a variety of organic and inorganic chemicals, but the predominant components are polysaccharides (Singha, 2012). Many methanotrophic microorganisms possess the ability to EPS secretion (Hilger et al. 1999; Wilshusen et al. 2004a). Moreover, other types of bacteria, such as *Pseudomonas*, possess the same ability (Kachlany et al. 2001). Intensive creation of EPS is probably caused by long, excessive methane loading rate (Huber-Humer et al. 2008). Visible increment of EPS was observed in the biofilters with forced flow, where the gas was led in at a constant overpressure (Wilshusen et al. 2004a, Haubrichs & Widmann, 2006). However, the phenomenon of EPS creation was not observed in the landfill covers, in which the gas migrated under its own pressure; whereas, the methane load, counted per surface area of the layer of the filter bed material, was variable in time. It depended on the difference of the atmospheric pressure and the pressure inside the landfill.

Besides the clogging of soil pores, due to production of exopolimeric substances, also mechanical and chemical destruction of the filter bed material are the reasons for the gas porosity decrease. These can lead to the increase of micropores share. Excess of micropores in the material increases water holding capacity, slowing down the rate of contaminants diffusion to the bacteria cells. What is more, the filter bed settlement is the effect of changes in the structure of the bed material, occurring under the influence of the mineralisation process and gravitational compaction of the material. Although there are no specific guidelines of the value of this parameter, which could be considered critical to biofilters; in a short time, noticeable large fluctuations in the height of the bed should still constitute a requirements for further and more detailed research.

With the decrease of the porosity, and increase in moisture content, the gas permeability of the material decreases. This results in a higher drop of a pressure, measured on a particular section of the way of gas flow, inside the biofilter. The maximum acceptable pressure drop in biofilters working at industrial scale is 1500 Pa m^{-1} of the bed (Estrada et al. 2012). This parameter is particularly important in the case of actively vented biofilters because an increase in pressure drop requires more blower power.

Moreover; a decrease in the pH value with time, resulting from the accumulation of acidic metabolic products as well as chemical reactions occurring in the material of the bed, may be the cause of decrease in the microbial activity in the bed material. In order to prevent the pH drop, buffering material, such as $CaCO_3$, may be added to the filter media.

Difficulties in the transport of pollutants into the microbial cells resulting from the high hydrophobicity of many organic compounds are the problems that concern both the biocovers and the biofilters. One of the tested solutions of this problem is the use of surfactants, in order to increase the solubility of degraded compounds. Ramirez et al. (2012) studied the effect of non-ionic surfactants (3 polyoxyethylenes (Brijs) and 3 mono polyoxyethylenesorbitans (Tweens) on the biofiltration of methane. They added surfactants to the nutrient solution at a concentration of 0.5% (by weight). Biofilters supported with surfactants showed higher methane removal efficiency; the RE value were: 38%–46% in biofilters with Brijs, 43–48% with Tweens, and 35% in biofilter without surfactant (35%). In biological systems using activated sludge, it is possible to use oil phase to incorporate hydrophobic microorganisms to the process. The effectiveness of this type of solution was examined by Muñoz et al. (2013) in two-phase (water and silicon oil) stirred tank reactor which was used for hexan removal.

Furthermore, uneven distribution, in space and time, of input of pollutants into the bed may pose another problem. This can lead to exceed the permissible values of loading rate of contaminants and decrease the efficiency of the process. Too high loading rate of pollutants, especially moderately-soluble VOCs, may inhibit the process due to the toxic influence to microorganisms. Moreover, too high value of gas loading rate leads to a high rate of gas flow through the filter bed causing the displace air from the soil pores (Gebert & Grögröft, 2006b). These problems can be mitigated in biofilters with controlled gas supply.

Sensibility of biological methods (except for closed biofilters) to climatic conditions is their major disadvantage. Seasonal changes in air temperature (Zeiss, 2006), and precipitation lead to changes in the temperature and the humidity of the bed. This problem concerns the temperate and boreal climate zones. The problem can be solved by introducing the temperature control (heat exchangers) and humidity (sprinkler system or irrigation reservoir of gas stream) which is associated with high operating costs. What is more, the disadvantage is also the fact that biofiltration is not able to ensure the removal of all of the compounds contained in LFG. Some compounds, which are recalcitrant to biologically-mediated degradation (e.g. certain chlorinated compounds), will not be converted to harmless products.

5.10   SUMMARY

The usage of microorganisms inhabiting the porous materials for mitigation of landfill gas emission is a solution for the future. In Europe, the need to replace previously used methods of thermal gas utilization leading to the production of energy with other methods is favoured. It is due to the gradual reduction of landfill gas emissions, which will cause by decrease in the deposition of organic matter in MSW landfills. In this context; biofiltration, relatively cheap, simple and not required complicated handling method demonstrates an enormous potential. It offers the opportunity to use the locally available soil or compost materials produced from organic waste, providing at the same time further opportunity for their utilisation.

The fundament for the effective operation of biofilters is to provide optimal conditions for the growth of methanotrophic microorganisms. This requires a proper selection of the bed material, maintenance of the adequate temperature of the process, and ensuring appropriate technological parameters. While choosing the bed, its general porosity should be taken into account. What is more, the participation of macro and micro pores, which determines the proper proportions of water and air in the material, should also be considered. Additionally, it is important to provide specific surface area, which has an influence on the conditions of the bacteria attachment on the material. The filter material should have low susceptibility to sedimentation because this parameter affects the time of the effective work of the biofilter. Apart from the physical parameters, the filter bed material should be characterized by appropriate chemical properties. It is necessary to assess the pH value, C: N ratio and the chemical stability of the material, which are assessed by the respirometric test. Stability has the influence on aerobic conditions in the bed. Oxygen uptake in the mineralization process of organic matter, which is contained in the filter material, reduces the availability of oxygen for the bacteria oxidising the compounds contained in the landfill gas. In order to create good conditions for the growth of bacteria with a high oxidizing capacity, it is necessary to ensure that the material does not contain high concentrations of copper, which inhibits the production of ions; and ammonium ions, which inhibit the oxidation of certain organic compounds blocking the activity of methane monooxygenase.

Large selection of technical and technological solutions of the process allows for the use of biological methods in landfills at different stages of life, with different structures and located in different environmental conditions.

At this level of recognition of mechanisms and determinants of processes of microbial removal of pollutants, it seems that one of the directions of further development of innovative biofilters for landfill gas mixture containing the methane and toxic components such as VOCs is the enhancement of the transfer of removed components to the microorganism cells.

REFERENCES

Abell, G.C., Stralis-Pavese, N., Sessitsch, A. & Bodrossy, L. 2009. Grazing affects methanotroph activity and diversity in an alpine meadow soil. *Environmental Microbiology Reports* 1(5), 457–465.
Abichou, T., Chanton, J., Powels, D., 2006b, Field Performance of biocells, biocovers, and biofilters to mitigate greenhouse gas emission from landfills. Annual Report of Research Project Funded by Florida Center for Solid and Hazardous Waste Management (www.floridacenter.org).
Abichou, T., Chanton, J., Powelson, D., Fleiger, J., Escoriaza, S., Yuan, L., Stern, J. 2006a. Methane flux and oxidation at two types of intermediate landfill cover. *Waste Manage.* 26: 1305–1312.
Ait-Benichou, S., Jugnia, L.B., Greer, C.W., and Cabral, A.R. (2009). Methanotrophs and methanotrophic activity in engineered landfill biocovers. *Waste Manage.* 29, 2509.
Albanna, M., Fernandes, F., Warith, M. 2007. Methane oxidation in landfill cover soil: the combined effects of moisture content, nutrient addition, and cover thickness. *J. Environ. Eng. Sci.* 6: 191–2000.
Alvarez, P.J. & Vogel, T.M. 1991. Substrate interactions of benzene, toluene, and *para*-xylene during microbial degradation by pure cultures and mixed culture aquifer slurries. *Appl Environ Microbiol.* 57(10): 2981–2985.
Arif, M.A.S., Houwen, F., Verstraete, W. 1996. Agricultural factors affecting methane oxidation in arable soil. *Biology and Fertility of Soils* 21(1–2): 95–102.
Arvin, E., Jensen, B.K., Gundersen, A.T. 1989, Substrate interactions during aerobic biodegradation of benzene. *Appl. Environ. Microbiol.* 55: 3221–3225.
Assinder, S. J., Williams, P. A. 1990. The TOL plasmids: determinants of the catabolism of toluene and the xylenes. *Adv. Microb. Physiol.* 31: 1–69.
Auman, A.J., Stolyar, S., Costello, A.M., Lidstrom, M.E. 2000. Molecular characterization of methanotrophic isolates from freshwater lake sediment. *Appl. Environ. Microbiol.* 66: 5259–5266.
Bader, R, Leisinger, T. 1994. Isolation and characterization of the *Methylophilus* sp. strain DM11 gene encoding dichloromethane dehalogenase/glutathione S-transferase. *J Bacteriol.* 176(12): 3466–3473.
Barbieri, P., Palladino, L., Gennaro, P.D., Galli, E. 1993. Alternative pathways for *o*-xylene or *m*-xylene and *p*-xylene degradation in a *Pseudomonas stutzeri* strain. *Biodegradation* 4 (2): 71–80.
Barlaz, M.A., Green, R.B., Chanton, J.P., Goldsmith, C.D., Hater, G.R. 2004. Evaluation of a biologically active cover for mitigation of landfill gas emissions. *Environ. Sci. Technol.* 38(18): 4891–4899.
Bender, M., Conrad, R. 1993. Kinetics of methane oxidation in oxic soils. *Chemosphere* 26(1–4): 687–696.
Bioaugmentation for Remediation of Chlorinated Solvents. Technology, Development, Status and Research Needs, ESTCP, October 2005 (http://www.clu-in.org/download/remed/Bioaug2005.pdf)

Boeckx, P., Van Cleemput, O. 1996. Methane oxidation in a neutral landfill cover Soil. Influence of moisture content, temperature, and nitrogen-turnover. *J. Environ. Qual.* 25: 178–183.

Boethling, R.S. 1986. Application of molecular topology to quantitative structure-biodegradability relationships. *Environmental Toxicology and Chemistry* 5(9): 797–806.

Bogner, J. & Spokas, K. 1993. Landfill CH_4: Rates, Fates, and Role In Global Carbon Cycle. *Chemosphere* 26(1–4): 369–386.

Bogner, J. Spokas, K. Chanton J. Powelson D., Fleiger, J. & Abichou T. Modeling landfill methane emissions from biocovers: a combined theoretical-empirical approach. In Proceedings Sardinia 2005 Tenth International Waste Management and Landfill Symposium, S. Margherita di Pula, Sardinia, Italy October 2005, Published by CISA, University of Cagliari, Sardinia.

Bogner, J.E., Chanton, J.P., Blake, D., Abichou, T., Powelson, D. 2010. Effectiveness of a Florida Landfill biocover for reduction of CH_4 and NMHC emissions. *Environ. Sci. Technol.* 44(4): 1197–1203.

Börjesson, G., Chanton, J., Svensson, B.H. 2001. Methane oxidation in two Swedish landfill covers measured with carbon-13 to carbon-12 isotope ratios. *J Environ Qual.* 30(2): 369–76.

Börjesson, G., Sundh, I, Svensson, B.H. 2004, Microbial oxidation of CH_4 in different temperatures in landfill cover soils, *FEMS Microbiol. Ecol.* 48: 305–312.

Bowman, J.P., Jiménez, L., Rosario, I., Hazen, T.C. Sayler, G.S. 1993. Characterization of the methanotrophic bacterial community present in a trichloroethylene-contaminated subsurface groundwater site. *Appl. Environ. Microbiol.* 59(8): 2380.

Bowman, J.P., 2006. The Methanotrophs—The Families Methylococcaceae and Methylocystaceae. *The Prokaryotes* 5: 266–289.

Brigmon, R.L., 2001. *Methanotrophic bacteria: Use in bioremediation*, Westinghouse Savannah River Co., Aiken, SC. Paper No: WSRC-MS-2001-00058. Feb 2001

Buchholz, L.A., Val Klump, J., Collins, M.L.P., Brantner, C.A. & Remsen, C.C. 1995. Activity of methanotrophic bacteria in Breen Bay sediments. *FEMS Microbiology Ecology* 16: 1–8.

Cabral, A.R., Moreira J.F.V., Jugnia, L.-B. 2010. Biocover Performance of Landfill Methane Oxidation: Experimental Results. *Journal of Environmental Engineering* 136(8): 785–793.

Cébron, A., Bodrossy, L., Chen, Y., Singer, A.C., Thompson, I.P., Prosser, J.I., Murrell, J.C. 2007. Identity of active methanotrophs in landfill cover soil as revealed by DNA-stable isotope probing. *FEMS Microbiol Ecol.* 62(1): 12–23.

Chan, A.S.K., Parkin, T.B. 2000. Evaluation of potential inhibitors of methanogenesis and methane oxidation in a landfill cover soil. *Soil Biol Biochem.* 32: 1581–1590.

Chanton, J., & Liptay, K. 2000. Seasonal variation in methane oxidation in a landfill cover soil as determined by an in situ stable isotope technique. *Global Biogeochem.* Cycles 14: 51.

Chanton, J.P, Powelson, D.K. Green, R.B. 2009. Methane oxidation in landfill cover soils, is a 10% default value reasonable?, *J Environ Qual,* 38:654–663.

Chen, Y., Dumont, M.G., Cebron, A., and Murrell, J.C. 200). Identification of active methanotrophs in a landfill cover soil through detection of expression of 16S rRNA and functional genes. *Environ. Microbiol.* 9: 2855.

Chen, Y., Crombie, A., Rahman, M.T., Dedysh, S.N., Liesack, W., Stott, M.B., Alam, M., Theisen, A.R., Murrell, J.C., Dunfield, P.F. 2010. Complete genome sequence of the aerobic facultative methanotroph *Methylocella silvestris* BL2. *J. Bacteriol.* 192(14): 3840–3841.

Chiemchaisri, W., Visvanathan, C., Wu, J.S. 2001. Effects of trace volatile organic compounds on methane oxidation. *Braz. Arch. Biol. Technol.* 44(2): 135–140.

Chistoserdova, L., Vorholt, J.A., and Lidstrom, M.E. 2005. A genomic view of methane oxidation by aerobic bacteria and anaerobic archaea. *Genome Biology* 6: 208.

Choi, E.J., Hyun, Jin, M., Lee S.H., Math, R.K., Madsen E.L. Jeon C.O. 2013. Comparative genomic analysis and benzene, toluene, ethylbenzene, and *o*-, *m*-, and *p*-xylene (BTEX) degradation pathways of *Pseudoxanthomonas spadix* BD-a59. *Appl. Environ. Microbiol.* 79 (2): 663–671.

Choi, E.N., Cho, M.C., Kim, Y., Kim, C-K., Lee, K., 2003. Expansion of growth substrate range in *Pseudomonas putida* F1 by mutations in both *cymR* and *tod*S, which recruit a ring-fission hydrolase CmtE and induce the tod catabolic operon, respectively. *Microbiology* 149: 795–805.

Christophersen, M., Linderød, L., Jensen, P.E., Kjeldsen, P. 2000. Methane oxidation at low temperatures in soil exposed to landfill gas. *J. Environ. Qual.* 29: 1989–1997.

Colby, J., Stirling, D.I., Dalton, H., 1997. The soluble methane mono-oxygenase *of Methylococcus capsulatus* (Bath). Its ability to oxygenate n-alkanes, n-alkenes, ethers, and alicyclic, aromatic and heterocyclic compounds. *Biochem J.* 165(2): 395–402.

Contin, M., Rizzardini, C.B., Catalano, L., De Nobili, M. 2012. Contamination by mercury affects methane oxidation capacity of aerobic arable soils. *Geoderma* 189–190: 250–256.

Costello, A.M., Auman, A.J., Macalady, J.L., Scow, K.M., Lidstrom, M.E. 2002. Estimation of methanotroph abundance in a freshwater lake sediment. *Environmental Microbiology* 4(8): 443–450.

Czepiel, P.M., Mosher, B., Crill, P.M. Harriss, R.C. 1996. Quantifying the effect of oxidation on landfill methane emissions. *Journal of Geophysical Research D: Atmospheres* 101(11): 16721–16729.

Dammann, B., Streese, J., Stegmann, R. 1999. Microbial oxidation of methane from landfills in biofilters. In T.H. Christensen, R. Cossu, R. Stegmann (eds.). *SARDINIA '99, In: Proceedings of the 7th International Landfill Symposium.* Vol. 2. CISA, Cagliari, Italy: 517–524.

De Visscher, A., Thomas, D., Boeckx, P., Van Cleemput, O. 1999. Methane oxidation in simulated landfill cover soil environments. *Environ. Sci. Technology* 33: 1854–1859.

Dedysh, S.N. & Dunfield, P.F. 2011. Facultative and obligate methanotrophs how to identify and differentiate them. *Methods Enzymol.* 495: 31–44.

Delhoménie, M-C., Bibeau, L., Gendron, J., Brzezinski, R., Heitz, M. 2001. Toluene removal by biofiltration: influence of the nitrogen concentration on operational parameters, *Ind. Eng. Chem. Res.* 40(23): 5405–5414.

Dever, S.A., Swarbrick, G.E., Stuetz, R.M. 2011. Passive drainage and biofiltration of landfill gas: results of Australian field trial. *Waste Manag.* 31(5): 1029–1048.

Devinny J.S., Deshusses, M.A., Webster T.S. 1999. *Biofiltration for Air Pollution Control.* CRC Press LLC.

Directive 1999/31/EC of 26 April 1999 on the landfill of waste L 182, 16 July 1999, pp. 1–19.

DiSpirito, A.A, Gulledge, J, Shiemke, A.K, Murrell, J.C, Lidstrom, M.E., Krema, C.L. 1991/1992. Trichloroethylene oxidation by the membrane-associated methane monooxygenase in type I, type II and type X methanotrophs. *Biodegradation* 2(3): 151–164.

Du Plessis, C.A., Strauss, J.M., Sebapalo, E.M.T. Riedel K.H.J. 2003. Empirical model for methane oxidation using a composted pine biofilter. *Fuel* 82: 1359–1365.

Dubey, S. K. & Singh, J. S. 2000. Spatio-temporal variation and effect of urea fertilization on methanotrophs in a tropical dryland rice field. *Soil Biology and Biochemistry* 32(4): 521–526.

Duetz, W.A., de Jong, C., Williams, P.A., van Andel, J.G. 1994. Competition in chemostat culture between pseudomonas strains that use different pathways for the degradation of toluene. *Applied and Environmental Microbiology* 60(8): 2858–2863.

Dunfield, P.F., Yuryev, A., Senin, P., Smirnova, A.V., Stott, M.B., Hou, S.B., Ly, B., Saw, J.H., Zhou, Z.M., Ren, Y., et al. 2007. Methane oxidation by an extremely acidophilic bacterium of the phylum Verrucomicrobia. *Nature* 450, 879–818.

Eaton, R.W. 1997. p-Cymene catabolic pathway in *Pseudomonas putida* F1: cloning and characterization of DNA encoding conversion of p-cymene to p-cumate. *J. Bacteriol.* 179(10): 3171–3180.

Einola, J-K., Kettunen, R.H., Rintala, J.A. 2007. Responses of methane oxidation to temperature and water content in cover soil of a boreal landfill. *Soil Biol. Biochem.* 39(5): 1156–1164.

Einola, J., Sormunen, K., Lensu, A., Leiskallio, A., Ettala, M., and Rintala, J. 2009. Methane oxidation at a surface-sealed boreal landfill. *Waste Management* 29: 2105–2120.

Emanuelsson, M.A.E., Osuna, M.B., Ferreira, J.R.M. Castro, P.M.L. 2009. Isolation of a *Xanthobacter* sp. degrading dichloromethane and characterization of the gene involved in the degradation. *Biodegradation* 20(2): 235–244.

Ensley, B.D. 1991. Biochemical diversity of trichloroethylene metabolism. *Annu Rev Microbiol.* 45: 283–299.

EPA ChSP. Chemical specific parameters (http://www.epa.gov/superfund/health/conmedia/soil/pdfs/part_5.pdf).

Estrada, J.M., Kraakman, N.J.R., Lebrero, R. Muñoz, R. 2012. A sensitivity analysis of process design parameters, commodity prices and robustness on the economics of odour abatement technologies. *Biotechnology Advances* 30(6): 1354–1363.

Fang, J., Barcelona, M.J, Alvarez, P.J. 2000. Phospholipid compositional changes of five pseudomonad archetypes grown with and without toluene. *Appl Microbiol Biotechnol.* 54(3): 382–389.

Fogel, M.M., Taddeo, A.R., Fogel, S. 1986. Biodegradation of chlorinated ethenes by a methane-utilizing mixed culture. *Appl Environ Microbiol.* 51(4): 720–724.

Frascari, D., Pinelli, D., Nocentini, M., Zannoni, A., Fedi, S., Baleani, E., Zannoni, D., Farneti, A., Battistelli, A. 2006. Long-term aerobic cometabolism of a chlorinated solvent mixture by vinyl chloride-, methane- and propane-utilizing biomasses. *J Hazard Mater.* 138(1): 29–39.

Fritsche, W. & Hofrichter, M. 2008. Aerobic degradation by microorganisms, in biotechnology: Environmental processes II, Vol.11b, Second Edition H.-J. Rehm & G. Reed (eds), Wiley-VCH Verlag GmbH, Weinheim, Germany.

Garcıa-Pena, I., Ortiz, I., Hernandez, S., Revah S., 2008. Biofiltration of BTEX by the fungus *Paecilomyces variotii, International Biodeterioration & Biodegradation* 62: 442–447.

Gąszczak, A., Szczyrba, E., Bartelmus, G., 2009. Kinetics studies of the biodegradation of volatile organic compounds in a batch reactor, *Proceedings of ECOpole* Vol. 3, No. 2; 295–300.

Gebert, J., Gröngröft, A., a& Miehlich, G. 2003. Kinetics of microbial landfill methane oxidation in biofilters. *Waste Manage.* 23: 609.

Gebert, J., Rachor, I., Bodrossy, L., 2009. Composition and activity of methane oxidizing communities in landfill covers. *Proceedings Sardinia 2009 Twelfth International Waste Management and Landfill Symposium* S. Margherita di Pula, Cagliari, Italy, CISA Publisher, Italy.

Gebert, J., Stralis-Pavese, N., Alawi, M., Bodrossy, L. 2008. Analysis of methanotrophic communities in landfill biofilters using diagnostic microarray. *Environ Microbiol.* 10(5): 1175–1188.

GESTIS substance database IFA (http://gestis-en.itrust.de/nxt/gateway.dll/gestis_en/010070.xml?f=templates$fn=default.htm$3.0).

Gibson, D.T, Parales, R.E. 2000. Aromatic hydrocarbon dioxygenases in environmental biotechnology. *Curr Opin Biotechnol.* 11: 236–243.

Gibson, D.T., S.M. Resnick, K. Lee, J.M. Brand, D.S. Torok, L.P. Wackett, M.J. Schocken, and B.E. Haigler. 1995. Desaturation, Dioxygenation, and Monooxygenation Reactions Catalyzed by Naphthalene Dioxygenase from *Pseudomonas* sp. Strain 9816-4. *J. Bacteriol.* 177: 2615–2621.

Gibson, D.T., J.R. Koch, and R.E. Kallio. 1968. oxidative degradation of aromatic hydrocarbons by microorganisms. I. Enzymatic formation of atechol from benzene. *Biochemistry* 7: 2653–2662.

Gilbert, B., McDonald, I.R., Finch, R., Stafford, G.P., Nielsen, A.K., and Murrell, J.C. 2000. Molecular analysis of the pmo (particulate methane monooxygenase) operons from two type II methanotrophs. *Appl. Environ. Microbiol.* 66: 966.

Grosse, S., Laramee, L. Wendlandt, K.D., McDonald, I.R., Miguez, C.B. and Kleber, H.P. 1999. Purification and characterization of the soluble methane monooxygenase of the type II methanotrophic bacterium *Methylocystis* sp. strain WI 14. *Appl Environ Microbiol.* 65(9): 3929–3935.

Gülensoy, N. Alvarez, P.J.J. 1999. Diversity and correlation of specific aromatic hydrocarbon biodegradation capabilities. 10: *Biodegradation* 331–340.

Guo, YH. 1990. Microbial kinetics of *Pseudomonas* sp. straub DM1 during dichloromethane biodegradation. *Chin J Biotechnol.* 6(1): 75–85.

Guzik U., Hupert-Kocurek K., & Wojcieszyńska D., 2013. Intradiol Dioxygenases — The Key Enzymes in Xenobiotics Degradation. In R. Chamy & F. Rosenkranz (eds) *Agricultural and Biological Sciences Biodegradation of Hazardous and Special Products.*

Halet, D., Boon, N., Verstraete, W. 2006. Community dynamics of methanotrophic bacteria during composting of organic matter. *J Biosci Bioeng.* 101(4): 297–302.

Han, X., Scott, A.C., Fedorak, P.M., Bataineh, M. & Martin, J.W. 2008. Influence of molecular structure on the biodegradability of naphthenic acids, *Environ. Sci. Technol.* 42(4): 1290–1295.

Hanson, R.S., & Hanson, T.E. 1996. Methanotrophic bacteria. *Microbiol. Rev.* 2: 439.

Haubrichs, R. & Widmann, R. 2006. Evaluation of aerated biofilter systems for microbial methane oxidation of poor landfill gas. *Waste Management* 26: 673–674.

Haug, R.T. 1993. *The Practical Handbook of Compost Engineering.* Boca Raton, FL: Lewis Publishers.

Haughey, R.D., 2001. Landfill alternative daily cover: conserving airspace and reducing landfill operating cost. *J. Waste Manag. Res.* 19: 89–95.

He, R., Ruan, A., Jiang, C., and Shen, D. 2008. Responses of oxidation rate and microbial communities to methane in simulated landfill cover soil microcosms. *Bioresour. Technol.* 99: 7192.

Henckel, T., Roslev, P., and Conrad, R. 2000. Effects of O_2 and CH_4 on presence and activity of the indigenous methanotrophic community in rice field soil. *Environ Microbiol.* 2: 666–679.

Hesselsoe, M, Boysen, S, Iversen, N., Jørgensen, L., Murrell, J.C., McDonald, I., Radajewski, S., Thestrup, H, Roslev, P. 2005. Degradation of organic pollutants by methane grown microbial consortia. *Biodegradation* 16(5): 435–448.

Higgins, I.J, Best, D.J., Hammond, R.C., Scott, D. 1981. Methane-oxidizing microorganisms microbiological reviews 45(4): 556–590.

Hilger, H., Bogner, J., Adams, B. & Hamm, J. 2007. Bio-tarp: developing a methanotrophic alternative daily cover to reduce landfill methane emissions. In: Eleventh International Waste Management and Landfill Symposium, 1–5 October 2007, S. Margherita di Pula, Cagliari, Italy.

Hoeks, J. 1972. *Effect of leaking natural gas on soil and vegetation in urban areas.* Agricultural Research Reports 778, Centre for Agricultural Publishing and Documentation (Pudoc), Wageningen, Netherlands.

Horz, H.-P., Raghubanshi, A.S., Heyer, J., Kammann, C., Conrad, R., and Dunfield, P.F. 2002. Activity and community structure of methane-oxidising bacteria in a wet meadow soil. *FEMS Microbiol. Ecol.* 41: 247–257.

Hršak, D. & Begonja, A., 2000. Possible interactions within a methanotrophic-heterotrophic ground-water community able to transform linear alkylbenzenesulfonates. *Appl. Environ. Microbiol.* 66 (10): 4433–4439.

Huang, Q., Zhang, Q., Cicek, N., Mann, D. 2011. Biofilter: a promising tool for mitigating methane emission from manure storage. *Journal of Arid Land* 3(1): 61–70.

Huber-Humer, M., Gebert, J., Hilger, H. 2008. Biotic systems to mitigate landfill methane emissions. *Waste Manage. Res.* 26: 33–46.

Huber-Humer, M., Röder, S., Lechner, P. 2009. Approaches to assess biocover performance on landfills, *Waste Manage.* 29: 2092–2104.

Humer, M. & Lechner, P., 1999. Alterative approach to the elimination of greenhouse gases. *Waste Manage. Res.* 17(6): 443–452.

Hütsch, B.W. 1998. Methane oxidation in arable soil as inhibited by ammonium, nitrite, and organic manure with respect to soil pH. *Biology and Fertility of Soils* 28(1): 27–35.

ICSC Database, International Chemical Safety Cards, Center For Disease Control And Prevention (http://www.cdc.gov/niosh/ipcs/icstart.html).

IPCC 2007. Forster, P., V. Ramaswamy, P. Artaxo, T. Berntsen, R. Betts, D.W. Fahey, J. Haywood, J. Lean, D.C. Lowe, G. Myhre, J. Nganga, R. Prinn, G. Raga, M. Schulz and R. Van Dorland, 2007: Changes in Atmospheric Constituents and in Radiative Forcing. In: Climate Change 2007: The Physical Science Basis. Contribution of Working Group I to the Fourth Assessment Report of the Intergovernmental Panel on Climate Change [Solomon, S., D. Qin, M. Manning, Z. Chen, M. Marquis, K.B. Averyt, M. Tignor and H.L. Miller (eds.)]. Cambridge University Press, Cambridge, United Kingdom and New York, NY, USA.

Islam, T, Jensen, S., Reigstad, L.J., Larsen, Ø., and Birkeland, N.K. 2008. Methane oxidation at 55°C and pH 2 by a thermoacidophilic bacterium belonging to the Verrucomicrobia phylum. *PNAS* 105: 300.

Jang, J.Y, Kim, D., Bae, H.W., Choi, K.Y, Chae, J.C., Zylstra, G.J., Kim, Y..M., Kim, E. 2005. Isolation and characterization of a *Rhodococcus* species strain able to grow on *ortho-* and *para*-xylene. *J Microbiol.* 43(4): 325–330.

Jones, H.A. &. Nedwell, D.B. 1993. Methane emission and methane oxidation in landfill cover soil. *FEMS Microbiology Ecology* 102(3–4): 185–195.

Jørgensen, C., Nielsen, B., Jensen, B.K., Mortensen, E. 1995. Transformation of o-xylene to o-methyl benzoic acid by a denitrifying enrichment culture using toluene as the primary substrate. *Biodegradation* 6: 141–146.

Jugnia, L.B., Ait-Benichou, S., Fortin, N., Cabral, A.R., and Greer, C.W. 2009. Diversity and dynamics of methanotrophs within an experimental landfill cover soil. *Soil Sci.Soc. Am. J.* 73, 1479.

Juhasz A.L. & Naidu R. 2000. Bioremediation of high molecular weight polycyclic aromatic hydrocarbons: a review of the microbial degradation of benzo[a]pyrene. *International Biodeterioration & Biodegradation* 45(1–2): 57–88.

Kachlany, S.C., Levery, S.B., Kim, J.S., Reuhs, B.L., Lion, L.W, Ghiorse, W.C. 2001. Structure and carbohydrate analysis of the exopolysaccharide capsule of *Pseudomonas putida* G7. *Environmental Microbiology* 3(12): 774–784.

Kallistova, A.Y., Kevbrina, M.V., Nekrasova, V.K., Shnyrev, N.A., Einola, J.K.M., Kulomaa, M.S., Rintala, J.A., & Nozhevnikova, A.N. 2007. Enumeration of methanotrophic bacteria in the cover soil of an aged municipal landfill. *Microb. Ecol.* 54: 637.

Kalyuzhnaya M.G., Khmelenina V.N., Suzina N.E., Lysenko A.M., and Trotsenko Y.A. 1999. New methanotrophic isolates from soda lakes of the southeastern Transbaikal region. *Microbiol.* 68: 677.

Khmelenina, V.N., Kalyuzhnaya, M.G., Starostina, N.G., Suuzina, N.E., & Trotsenko, Y.A. 1997. Isolation and characterization of halotolerant alkaliphilic methanotrophic bacteria from Tuva soda lakes. *Curr. Microbiol.* 35: 257.

Kightley, D., Nedwell, D.B. Cooper, M. 1995. Capacity for methane oxidation in landfill cover soils measured in laboratory-scale soil microcosms. *Appl. Environ. Microb.* 61(2): 592–601.

Kim, D., Kim, Y.-S., Kim, S.-K., Kim, S.W., Zylstra, G.J., Kim, Y.M., Kim, E. 2002. Monocyclic aromatic hydrocarbon degradation by *Rhodococcus* sp. strain DK17. *Appl. Environ. Microbiol.* 68(7): 3270–3278.

Kim, S.B., Park, C.H., Kim, D.J., Jury, W.A. 2003. Kinetics of benzene biodegradation by *Pseudomonas Aeruginosa*: Parameter estimation. *Environ Toxicol Chem.* 22(5): 1038–1045.

Kjeldsen P, Dalager, A., Broholm, K. 1997. Attenuation of methane and nonmethane organic compounds in landfill gas affected soils. *Journal of the Air & Waste Management Association* 47(12): 1268–1275.

Knief, C., Vanitchung, S., Harvey, N.W., Conrad, R., Dunfield, P.F., Chidthaisong, A. 2005. Diversity of methanotrophic bacteria in tropical upland soils under different land uses. *Applied and Environmental Microbiology* 71(7): 3826–3831.

Kolb, S., Knief, C., Stubner, S. Conrad, R. 2003. Quantitative detection of methanotrophs in soil by novel *pmoA*-Targeted Real-Time PCR assays. *Appl. Environ. Microbiol.* 69(5): 2423–2429.

Kong, J.Y., Su, Y., Zhang, Q.Q., Bai, Y., Xia, F.F., Fang, C.R., He, R. 2013. Vertical profiles of community and activity of methanotrophs in landfill cover soils of different age. *J Appl Microbiol.* 115(3): 756–65.

Kunze, M., Zerlin, K.F., Retzlaff, A., Pohl, J.O. Schmidt, E., Janssen, D.B., Vilchez-Vargas, R., Pieper D.H., Walter, Reineke W. 2009. Degradation of chloroaromatics by *Pseudomonas putida* GJ31: assembled route for chlorobenzen degradation encoded by clusters on plasmid KW1and the chromosome. *Microbiology* 155: 4069–4083.

Leahy, J.G, Batchelor, P.J, Morcomb, S.M. 2003. Evolution of the soluble diiron monooxygenases. *FEMS Microbiol Rev.* 27: 449–479.

Lee, E-H., Park, H., Cho, K-S. 2010. Characterization of methane, benzene and toluene-oxidizing consortia enriched rom landfill and riparian wetland soils, *Journal of Hazardous Materials* 184: 313–320.

Lee, E-H., Park, H., Kyung-Suk, Cho, K-S. 2013. Biodegradation of methane, benzene, and toluene by a consortium MBT14 enriched from a landfill cover soil. *Journal of Environmental Science and Health* Part A: *Toxic/Hazardous Substances and Environmental Engineering* 48(3): 273–278.

Lee, E.H, Park, H., Cho, K.S. 2011a. Effect of substrate interaction on oxidation of methane and benzene in enriched microbial consortia from landfill cover soil. *J Environ Sci Health a Tox Hazard Subst Environ Eng.* 46(9): 997–1007.

Lee, E-H., Yi, T., Moon, K.E., Park, H., Ryu, H.W., Cho, K.S. 2011b. Characterization of methane oxidation by a methanotroph isolated from a landfill cover soil, South Korea. *J Microbiol Biotechnol.* 21(7): 753–756.

Limtong, S., Srisuk, N., Yongmanitchai, W., Yurimoto, H., Nakase, T., 2008. *Ogataea chonburiensis* sp. nov. and *Ogataea nakhonphanomensis* sp. nov., thermotolerant, methylotrophic yeast species isolated in Thailand, and transfer of *Pichia siamensis* and *Pichia thermomethanolica* to the genus *Ogataea*. *International Journal of Systematic and Evolutionary Microbiology* 58(1): 302–307.

Little, C.D., Palumbo, A.V., Herbes, S.E., Lidstrom, M.E., Tyndall, R.L. &. Gilmer, P.L. 1988. Trichloroethylene biodegradation by a methane-oxidizing bacterium. *Appl. Environ. Microbiol.* 54: 951–956.

Liu, Y, Liu S-S., Cui S-H. & Cai S-X. 2003. A novel quantitative structure-biodegradability relationship (QSBR) of substituted benzenes based on MHDV description. *Journal of the Chinese Chemical Society* 50: 319–324.

Mac Donald, I.R., Hall, G.H., Pickup, R.W., Murrell, J.C., 1996. Methane oxidation potential and preliminary analysis of methanotrophs in blanket bog peat using molecular ecology techniques. *FEMS Microbiol. Ecol.* 21: 197–211.

Macalady J.L., McMillan A.M.S., Dickens A.F., Tyler S.C, Scow K.M. 2002. Population dynamics of type I and II methanotrophic bacteria in rice soils. *Environmental Microbiology* 4(3): 148–157.

Mars, A.E, Houwing, J., Dolfing, J. Janssen, D.B. 1996. Fed-Batch Culture by *Burkholderia cepacia* G4 in growth-limited degradation of toluene and trichloroethylene. *Appl. Environ. Microbiol.* 62(3): 886.

McDonald, I.R., Bodrossy, L, Chen, Y. & Murrell J.C. 2008. Molecular ecology techniques for the study of aerobic methanotrophs. *Appl. Environ. Microbiol.* 74(5): 1305–1315.

McDonald, I.R., Doronina, N.V., Trotsenko, Y.A., McAnullaand, C., Murrell J.C. 2001. *Hyphomicrobium chloromethanicum* sp. nov. and *Methylobacterium chloromethanicum* sp. nov., chloromethane-utilizing bacteria isolated from a polluted environment. *International Journal of Systematic and Evolutionary Microbiology* 51: 119–122.

McCarty, P.L., and Semprini, L., 1994, Ground-water treatment for chlorinated solvents, In Norris, R.D., Hinchee, R.E., Brown, R., McCarty, P.L, Semprini, L., Wilson, J.T., Kampbell, D.H., Reinhard, M., Bouwer, E.J., Borden, R.C., Vogel, T.M., Thomas, J.M., and Ward, C.H. (eds), *Handbook of Bioremediation*. Lewis Publishers, Boca Raton, FL: 87–116.

Melse, R.W. & Van der Werf, A.W. 2005. Biofiltration for mitigation of methane emission from animal husbandry. *Environ. Sci., Technol.* 39: 5460–5468.

Miller, L.G, Sasson, C., Oremland, R.S 1998. Difluoromethane, a new and improved inhibitor of methanotrophy. *Appl. Environ Microbiol* 64: 4357–4362.

Mohanty, S.R, Bharati K., Deepa N., Rao, V.R., Adhya, T.K. 2000. Influence of heavy metals on methane oxidation in tropical rice soils. *Ecotoxicol Environ Saf.*, 47(3): 277–84.

Mor, S., De Visscher, A., Ravindra, K., Dahiya, R.P. Chandra, A., Van Cleemput, O. 2006. Induction of enhanced methane oxidation in compost: Temperature and moisture response. *Waste Manage.* 26: 381–388.

Muñoz, R., Gan, E., Hernandez, M., Quijano, G. 2013. Hexane biodegradation in two-liquid phase bioreactors: high-performance operation based on the use of hydrophobic biomass. *Biochemical Engineering Journal* 70: 9–16.

Murrell, J.C., McDonald, I.R. & Gilbert, B. 2000. Regulation of expression of methane monooxygenases by copper ions. *Trends Microbiol.* 8: 221–225.

Murell J.C. 2010. The aerobic methane oxidizing bacteria (Methanotrophs). In *Handbook of hydrocarbon and lipid microbiology*: 1953–1966.

Nelson, M.J. K., Montgomery, S.O., Mahaffey, W.R & Pritchard, P.H. 1987. Biodegradation of trichloroethylene and involvement of an aromatic pathway. *Appl. Environ. Microbiol.* 53: 949–954.

Nikiema, J., Brzezinski, R. & Hertz, M. 2007. Elimination of methane generated from landfills by biofiltration: a review. *Reviews in Environmental Science and Biotechnology* 6(4): 261–284.

Nozhevnikova, A.N., Lifshitz, A.B., Lebedev, V.S., and Zavarzin, G.A. 1993. Emission of methane into the atmosphere from landfills in the former USSR. Chemosphere 26, 401

Olsen, R. H., J. J. Kukor, and B. Kaphammer. 1994. A novel toluene-3-monooxygenase pathway cloned from Pseudomonas pickettii PKO1. J. Bacteriol. 176: 3749–3756.

Onaca, C., Kieninger, M., Engesser, K-H, Altenbuchne, J. 2007. Degradation of alkyl methyl ketones by *Pseudomonas veronii. J. Bacteriol.* 189(10): 3759–3767.

Oremland, R.S., Culbertson, C.W. 1992. Importance of methane-oxidizing bacteria in the methane budget as revealed by the use of a specific inhibitor. *Nature* 356: 421–423.

Otenio, M.H., Lopes da Silva, M.T., Oliveira Marques M.L., Roseiro, J.C., Bidoia, E.D. 2005. Benzene, toluene and xylene biodegradation by Pseudomonas putida CCMI 852. *Braz. J. Microbiol.* 36 (3): 258–261.

Park, S., Lee, I., Cho, C., and Sung, K. (2008) Effects of earthworm cast and powdered activated carbon on methane removal capacity of landfill cover soils. *Chemosphere* 70: 1117.

Pawłowska M. 2008. Reduction of methane emission from landfills by its microbial oxidation in filter bed. In M. Pawlowska & L. Pawlowski (eds) *Management of Pollutant Emission from Landfills and Sludge,* Taylor & Francis Group, London; 3–20.

Pawłowska, M., Rożej A, Stępniewski W. 2011. The effect of bed properties on methane removal in an aerated biofilter—Model studies. *Waste Management* 31: 903–913.

Pawłowska, M., Czerwiński, J., Stępniewski, W. 2008. Variability of the of non-methane volatile organic compounds (NMVOC) composition in biogas from sorted and unsorted landfill material. *Arch. Environ. Prot.* 34(3): 287–307.

Pawłowska, M., Stępniewski, W., 2011. Enhancement of methane oxidation capacity in biofilter due to aeration—a laboratory study. In *Proceedings of 13 International Waste Management and Landfill Symposium.* San Margherita di Pula, Cagliari, Italy, 3–7 October 2011.

Pawłowska, M., Stępniewski, W., Czerwiński, J. 2003. The effect of texture on methane oxidation capacity on sand layer—a model laboratory study. In L. Pawłowski, M.R. Dudzińska & A. Pawłowski (eds), *Environmental Engineering Studies in Poland. Polish Research on the Way to EU.* Kluwer Academic/Plenum Publishers, New York, NY: 339–353.

Perdikea, K., Mehrotra, A.K., Hettiaratchi, J.P. 2008. Study of thin biocovers (TBC) for oxidizing uncaptured methane emissions in bioreactor landfills. *Waste Manag.* 28(8): 1364–1374.

Pol, A., Heijmans, K., Harhangi, H.R., Tedesco, D., Jetten, M.S., and Op den Camp H.J. 2007. Methanotrophy below pH 1 by a new Verrucomicrobia species. *Nature* 450: 874.

Powelson, D.K., Chanton, J., Abichou, T., Morales, J. 2006. Methane oxidization in water-spreading and compost biofilters. *Waste Manage. Res.* 24: 528–536.

Prior, S.D., Dalton, H. 1985. Acetylene as a suicide substrate and active site probe for methane monooxygenase from *Methylococcus capsulatus* (Bath). *FEMS Microbiol Lett* 29: 105–109.

Qi, B., Moe, W.M., Kinney, K.A. 2002. Biodegradation of volatile organic compounds by five fungal species. *Appl Microbiol Biotechnol.* 58(5): 684–689.

Ramirez, A.A., García-Aguilar, B.P., Jones, J.P., Heitz, M. 2012. Improvement of methane biofiltration by the addition of non-ionic surfactants to biofilters packed with inert materials, *Process Biochemistry* 47(1): 76–82.

Reay, D.S., Radajewski, S., Murrell, J.C. McNamara, N. & Nedwell, D.B. 2001. Effects of land-use on the activity and diversity of methane oxidizing bacteria in forest soils. *Soil Biol. Biochem.* 33: 1613–1623.

Reeburgh, W.S., Whalen, S.C., and Alperin, M.J. 1993. The role of methylotrophy in the global methane budget. In J.C. Murrell & D.P. Kelly (eds), Microbial Growth on C1 Compounds. Andover.

Ridgway, H., Safarik, F.J., Phipps, D., Carl, P. & Clark, D. 1990. Identification and catabolic activity of well-derived gasoline-degrading bacteria from a contaminated aquifer. *Appl. Environ. Microbiol.* 56: 3565–3575.

Ryu, H.W., Kim, S.J., Cho, K.S. 2010. Comparative studies on toluene removal and pressure drop in biofilters using different packing materials. *J Environ Biol.* 31(3): 315–318.

Sabourin, C.L., Carpenter, J.C., Leib, T.K. Spivack, J.L. 1996. Biodegradation of dimethylsilanediol in soils. *Appl Environ Microbiol.* 62(12): 4352–4360.

Sazinsky, M.H., Bard, J., Di Donato, A., & Lippard, S.L. 2004. Crystal Structure of the toluene/o-xylene monooxygenase hydroxylase from *Pseudomonas stutzeri* OX1 *J. Biol. Chem.* 279: 30600–30610.

Scheutz, C. Mosbæk, H., Kjeldsen, P. 2004. Attenuation of methane and volatile organic compounds in landfill soil covers. *J Environ Qual.* 33(1): 61–71.

Scheutz, C., Bogner, J., Chanton, J.P., Blake, D., Morcet, M., Aran, C., Kjeldsen P., 2008. Atmospheric emissions and attenuation of non-methane organic compounds in cover soils at a French landfill, *Waste Management* 28: 1892–1908.

Scheutz, C., Kjeldsen, P. 2003. Capacity for biodegradation of CFCs and HCFCs in a methane oxidative counter-gradient laboratory system simulating landfill soil covers, *Environ. Sci. Technol.* 37(22): 5143–5149.

Scheutz, C., Bogner, J., Chanton, J., Blake, D., Morcet, M., Kjeldsen, P. 2003. Comparative oxidation and net emissions of methane and selected non-methane organic compounds in landfill cover soils. *Environ sci. Technol.* 37(22): 5150–5158.

Scheutz, C., Mosbæk, H., Kjelsden, P. 2004. Attenuation of methane and volatile organic compounds in landfill soil covers., *Journal of environmental quality* 33: 61–71.

Scheutz, C., Kjeldsen, P. 2005. Biodegradation of Trace Gases in Simulated Landfill Soil Cover Systems, *J Air Waste Manag Assoc.* 55(7): 878–885.

Semrau, J.D., DiSpirito, A.A., Yoon S. 2010. Methanotrophs and copper. *FEMS Microbiol Rev* 34: 496–531.

Shields, M.S. & Montgomery S.O. 1989. Novel Pathway of Toluene Catabolism in the Trichloroethylene-Degrading Bacterium G4. *Appl. Environ. Microbiol.* 55: 1624–1629.

Singha, T.K. 2012. Microbial extracellular polymeric substances: Production, isolation and applications. *IOSR Journal of Pharmacy* 2(2): 276–281.

Sommer, C., Görisch, H. 1997 Enzymology of the degradation of (di) chlorobenzenes by *Xanthobacter flavus* 14p1. *Arch Microbiol.* 167(6): 384–391.

Stein, V.B. & Hettiaratchi, J.P.A. 2001. Methane oxidation in three alberta soils: Influence of soil parameters and methane flux rate. *Environ. Technol.* 22: 101–111.

Streese, J., Dammann, B., Stegmann, R. 2001. Eight Reduction of methane and trace gas emissions from former landfills in biofilters, *Proceedings Sardinia 2001, Eigth International Waste Management and Landfill Symposium.* S. Margherita di Pula, Cagliari, Italy 1–5 October 2001: 575–584.

Stralis-Pavese N., Bodrossy L., Reichenauer T.G., Weilharter A., and Sessitsch A. 2006. 16S rRNA based T-RFLP analysis of methane oxidising bacteria-Assessment, critical evaluation of methodology performance and application for landfill site cover soils. *Appl. Soil Ecol.* 31, 251.

Su, T-T., Lin, C-W., Yet-Po, I., Wu. C-H., 2012. Biodegradation of semiconductor volatile organic compounds by four novel bacterial strains: a kinetic analysis. *Bioprocess Biosyst Eng* 35: 1117–1124.

Sundh, I., Borga Ê,P., Nilsson, M., Svensson, B.H. 1995. Estimation of cell numbers of methanotrophic bacteria on boreal peatlands based on analysis of specific phospholipid fatty acids. *FEMS Microbiology Ecology* 18: 103–112.

Takacs-Novak, K. 2012. Physicochemical profiling in drug research and development. In Z. Mandic (ed) *Physico Chemical Methods in Drug Discovery and Development.* IAPC Publishing.

Theisen, A.R., Ali, M.H. Radajewski, S., Dumont, M.G., Dunfield, P.F., McDonald, I.R., Dedysh, S.N., Miguez, C.B., & Murrell J.C. 2005. Regulation of methane oxidation in the facultative methanotroph *Methylocella silvestris* BL2. *Mol. Microbiol.* 58: 682–692.

Trotsenko, Y.A. & Stępniewska, Z., 2012. *Biologia i biotechnologia aerobowych metylotrofów (Biology and biotechnology of aerobic methylotrophs)*, Wydawnictwo KUL, Lublin.

Trotsenko, Y.A., & Khmelenina, V.N., 2002. Biology of extremophilic and extremotolerant methanotrophs *Archives of Microbiology* 177(2): 123–131.

Tsubota, J., Eshinimaev, B.T., Khmelenina, V.N., and Trotsenko, Y.A. (2005). *Methylothermus thermalis* gen. nov., sp nov., a novel moderately thermophilic obligate methanotroph from a hot spring in Japan. *International Journal of Systematic and Evolutionary Microbiology* 55: 1877–1884.

Urmann, K.Æ., Martin, H. Schroth, Æ., Zeyer, J. 2008. Recovery of in-situ methanotrophic activity following acetylene inhibition. *Biogeochemistry* 89: 347–355.

Visvanathan, C., Pokhrel, D., Cheimchaisri, W., Hettiaratchi, J.P.A., Wu, J.S. 1999. Methanotrophic activities in tropical landfill cover soils: effect of temperature, moisture content and methane concentration, *Waste Manage. Res.* 17(4): 313–323.

Walkiewicz, A., Bulak, P., Brzezińska, M., Włodarczyk, T., Polakowski C., 2012. Kinetics of methane oxidation in selected mineral soils. *Int. Agrophys.* 26: 401–406.

Wang, Z.P., Ineson, P. 2003. Methane oxidation in a temperate coniferous forest soil: effects of inorganic N. *Soil Biology and Biochemistry* 35(3): 427–433.

Wang, H., Einola, J., Heinonen, M., Kulomaa, M., and Rintala, J. 2008. Group specific quantification of methanotrophs in landfill gas-purged laboratory biofilters by tyramide signal amplification-fluorescence in situ hybridization. *Bioresour. Technol.* 99: 6426.

Whalen, S.C., Reeburgh, W. S., Sandbeck, K.A. 1990. Rapid methane oxidation in a landfill cover soil, *Appl. Environ. Microb.* 56(11): 3405–3411.

Whited, G.M., & Gibson, D.T. 1991. Separation and partial characterization of the enzymes of the toluene-4-monooxygenase catabolic pathway in Pseudomonas mendocina KR1. *J. Bacteriol.* 173: 3017–3020.

Wilshusen, J.H., Hettiaratchi, J.P.A., De Visscher, A., Saint-Fort, R. 2004a. Methane oxidation and formation of EPS in compost: effect of oxygen concentration. *Environ. Pollut.* 129: 305–314.

Wilshusen, J.H., Hettiaratchi, J.P.A, Stein, V.B., 2004b, Long-term behavior of passively aerated compost methanotrophic biofilter columns. *Waste Manage.*, 24(7): 643–653.

Wise, M.G., McArthur, J.V., & Shimkets, L.J. 1990. Methanotroph diversity in landfill soil: isolation of novel type I and type II methanotrophs whose presence was suggested by culture-independent 16S ribosomal DNA analysis. *Appl. Environ. Microbiol.* 65: 4887.

Wolf, H.J. & Hanson, R.S. 1979. Isolation and characterization of methane-utilizing yeasts. *Journal of General Microbiology.* 114: 187–194.

Wolf, H.J., Christiansen, M. & Hanson, R.S. 1980. Ultrastructure of methanotrophic yeasts. *J Bacteriol.* 141(3): 1340–1349.

Wolf, H.J., Hanson, R.S. 1980. Identification of methane-utilising yeasts. *FEMS (Fed. Eur. Microbiol. Soc.) Microbiol. Lett.* 7: 177–179.

Yan, T., Ye, Q., Zhou, J., Zhang, C.L. 2006. Diversity of functional genes for methanotrophs in sediments associated with gas hydrates and hydrocarbon seeps in the Gulf of Mexico. *FEMS Microbiol Ecol.* 57(2): 251–259.

Yen, K.M., Karl, M.R., Blatt, L.M., Simon, M.J., Winter, R.B., Fausset, P.R., Lu, H.S., Harcourt, A.A. & Chen, K.K., 1991. Cloning and characterization of a Pseudomonas mendocina KR1 gene cluster encoding toluene-4-monooxygenase. *J Bacteriol.* 173(17): 5315–5327.

Yoon, S., Semrau, JD. 2008. Measurement and modelling of multiple substrate oxidation by methanotrophs at 20 degrees C. *FEMS Microbiol Lett.* 287(2): 156–162.

Yu, S.S.F., Chen, K.H.C., Tseng, M.Y.H., Wang, Y.S., Tseng, C.F., Chen, Y., Huang, D., and Chan, S. 2003. Production of high-quality particulate methane monooxygenase in high yields from Methylococcus capsulatus (Bath) with a hollow-fiber membrane bioreactor. *J. Bacteriol.* 185, 5915.

Zeiss, C.A. 2006, Accelerated methane oxidation cover systems to reduce greenhouse gas emission from MSW landfills in cold-semi arid regions *Water Air Soil Poll.* 176: 285–306.

Zhang, L.L., Leng, S.Q., Zhu, R.Y., Chen, J.M. 2011. Degradation of chlorobenzene by strain *Ralstonia pickettii* L2 isolated from a biotrickling filter treating a chlorobenzene-contaminated gas stream. *Appl Microbiol Biotechnol.* 91(2): 407–415.

Zilli M., Del Borghi A., & Converti A., 2000. Toluene vapour removal in a laboratory-scale biofilter. *Appl Microbiol Biotechnol.* 54(2): 248–254.

Zylstra, G .J., McCombie, W.R, Gibson, D.T., Finette, B.A. 1988. Toluene degradation by *Pseudomonas putida* F1: genetic organization of the *tod* operon. *Appl Environ Microbiol.* 54: 1498–1503.

Zylstra, G.J. 1994. Molecular analysis of aromatic hydrocarbon degradation. In S.J. Garte (ed) *Molecular Environmental Biology*. Lewis Publishers: Boca Raton, FL: 83–115.

Subject index